D0185704

Spon's
Landscape Contract
Handbook

Other titles from E & FN Spon

Amenity Landscape Management: A resources handbook
R. Cobham

Beazley's Design and Detail of the Space Between Buildings
A. Pinder and A. Pinder

Construction Contracts: Law and management
J. Murdoch and W. Hughes

Countryside Recreation, Access and Land Use Planning
N. Curry

Countryside Management
P. Bromley

Countryside Recreation Sites: A handbook for managers
P. Bromley

Elements of Visual Design in the Landscape
S. Bell

Environmental Planning for Site Development
A.R. Beer

Fungal Diseases of Amenity Turf Grasses
J. Drew Smith, N. Jackson and R. Woolhouse

The Golf Course
F.W. Hawtree

Grounds Maintenance: A contractor's guide to competitive tendering
P. Sayers

A Handbook of Segmental Paving
A.A. Lilley

Project Management Demystified: Today's tools and techniques
G. Reiss

Recreational Land Management
W. Seabrooke and C.W.N. Miles

Spon's Grounds Maintenance Contract Handbook
R.M. Chadwick

Spon's Landscape and External Works Price Book
Derek Lovejoy and Partners and Davis Langdon & Everest

Spon's Landscape Handbook: Third edition
Derek Lovejoy and Partners

Spon's Quarry Guide
D.I.E. Jones, H. Gill and J.L. Watson

Tree Form, Size and Colour
B.J. Gruffydd

Trees in the Urban Landscape: Principles and practice
A. Bradshaw, B. Hunt and T. Walmsley

Effective Speaking: Communicating in speech
C. Turk

Effective Writing: Improving scientific, technical and business communication
C. Turk and J. Kirkman

Good Style: Writing for science and technology
J. Kirkman

For more information on these and other titles please contact:
The Promotion Department, E & FN Spon, 2–6 Boundary Row, London, SE1 8HN. Telephone 0171–865 0066.

Spon's Landscape Contract Handbook

A guide to good practice and procedures in the management of lump sum landscape contracts

SECOND EDITION

HUGH CLAMP

Manning Clamp
Chartered Architects and Landscape Consultants

E & FN SPON
An Imprint of Chapman & Hall
London · Glasgow · Weinheim · New York · Tokyo · Melbourne · Madras

Published by E & FN Spon, an imprint of Chapman & Hall, 2–6 Boundary Row, London SE1 8HN, UK

Chapman & Hall, 2–6 Boundary Row, London SE1 8HN, UK

Blackie Academic & Professional, Wester Cleddens Road, Bishopbriggs, Glasgow G64 2NZ, UK

Chapman & Hall GmbH, Pappelallee 3, 69469 Weinheim, Germany

Chapman & Hall USA, 115 Fifth Avenue, New York, NY 10003, USA

Chapman & Hall Japan, ITP-Japan, Kyowa Building, 3F, 2-2-1 Hirakawacho, Chiyoda-ku, Tokyo 102, Japan

Chapman & Hall Australia, 102 Dodds Street, South Melbourne, Victoria 3205, Australia

Chapman & Hall India, R. Seshadri, 32 Second Main Road, CIT East, Madras 600 035, India

First edition 1986
Second edition 1995

Typeset in 10/12 pt Times New Roman by Pure Tech Corporation, Pondicherry, India
Printed in England by Clays Ltd, St Ives plc, Bungay, Suffolk

ISBN 0 419 18300 0

A catalogue record for this book is available from the British Library

Library of Congress Catalog Card Number: 95-68507

∞ Printed on permanent acid-free text paper, manufactured in accordance with ANSI/NISO Z39.48-1992 and ANSI/NISO Z39.48–1984 (Permanence of Paper).

From the reviews of the first edition:

- It is a book all practices should own and whose advice all practitioners should possess.

  Landscape Design

- . . . it constitutes an invaluable aid to students, practitioners and contractors, being written in a readable, unstuffy style, while achieving the author's stated aim of highlighting the main elements of landscape contractual procedures.

  Landscape and Urban Planning

- Sets out the best practices and procedures for all the parties concerned in landscape contracts . . . for anyone professionally involved in the landscape industry.

  Chartered Surveyor Weekly

- Everything a rural planning office will need to know about how a landscape architect draws up a landscape contract.

  International Yearbook of Rural Planning

- . . . the contents include a wealth of information of use to all those involved in the landscape industry

  Turf Management

- Invaluable for everyone professionally involved in the landscape industry . . . The author's experience of private practice and contract drafting makes the book a good mixture of practical experience and an industry-wide overview.

  Aslib Book List

'When you have to draw up a will or a contract, you have to choose your words well. You have to look into the future – envisage all the contingencies that may come to pass – and then use words to provide for them. On the words you use, your clients' future may depend.'

Lord Denning, *The Discipline of Law*,
Butterworths (1979), London

Contents

Preface to the second edition

Since the first edition was published in 1986 there have been only three new judgements providing further case law on the JCLI Form two, *Oram Builders v. M.J. Pemberton* and *Crestav Ltd. v. Carr* in 1987 concerning the powers of the arbitrator and one in 1991 concerning insurance against storm damage.

There have however been no less than seven revisions necessary to the Contract itself to keep it up to date in respect of current legislation, practice and procedures within the industry, the range of types of contract under which landscape work is procured has been extended by the increased use of Design/Build and Management methods. The work of the Coordinating Committee on Project Information (CCPI) culminated in the publication in 1987 of the final version of the Common Agreement properly coordinating for the first time drawings specification and bills of quantity, the seventh edition (SMM7) providing for detailed landscape quantities being issued at the same time.

The Local Government Act of 1988 introduced the obligation of local authorities to invite competitive tenders for landscape maintenance and the introduction of BS 5750 introduced the concept of Quality Assurance to the landscape industry.

Finally the opportunity has been taken to re-group certain sections of the book into a more orderly sequence and to redraft others to make their meaning more readily apparent to the reader.

Hugh Clamp
OBE, VRD, FCIArb, FRIBA, PPLA

Preface to the first edition

This manual is intended for all who are involved in the landscape industry: landscape architects, local and county authority parks directors and managers, landscape contractors, and the nurserymen who supply them. Most will know in detail their own duties and responsibilities but few will know in equal detail the duties and responsibilities of those with whom they come in daily contact.

It is hoped that by explaining the best practice and procedures of all the parties concerned in the satisfactory completion of landscape contracts – large and small, in public and in private sectors – each will be enabled to play his or her own part as effectively as possible.

This manual is particularly intended to ensure that plants which are tended with loving care over several years in the nursery are handled with equal care in transit to their new location, and are then planted and looked after in such a way that they thrive and grow unhindered in the years to come. It requires only one careless slip by any one of the many people responsible in the chain for the plants to become irreparably damaged or even killed as they are passed from one to the other.

The Joint Council for the Landscape Industry (JCLI) – comprising senior representatives of the Landscape Institute (LI), the British Association of Landscape Industries (BALI), the National Farmers Union Horticulture Division (NFU), Arboricultural Association (AA), the Institute of Leisure and Amenity Management (ILAM) and the Horticultural Trades Association (HTA) – was set up in 1974 under the chairmanship of J. Bernhard. It meets regularly throughout the year to ensure that best practices are achieved, as does its sister committee: the Joint Liaison Committee for Plants Supplies and Establishment (CPSE). It has to be admitted that in spite of much effort their recommendations do not always reach those for whom they are intended; it is hoped that by publishing them in the form of this book they will reach a wider audience.

The success of any landscape project depends on many things but few are more important than the successful relationship between the grower, the contractor and the manager or designer, and it is hoped to attempt to build bridges between them in the following pages.

Although primarily intended for the practitioners, the needs of the educationalist must not be overlooked and if the advice given is of assistance to the staff and students in the many schools of landscape architecture and horticultural colleges throughout the country then this might alone justify its production.

Acknowledgements

This book would not have been written but for the encouragement over the years of the many members of the Landscape Industry contributing to the continuing work of the Joint Council of Landscape Industries.

In acknowledging their help I must emphasize that all opinions are my own, written as a practising landscape architect and not a legal expert; they do not necessarily represent the views of any of the professional institutes to which I belong, nor of any of the committees in which I am involved.

Extracts from British Standards are reproduced with permission of the British Standards Institution. Complete copies can be obtained from BSI at Linford Wood, Milton Keynes, MK14 6LE.

The JCT Standard Form of Tender by Nominated Supplier is reproduced in Appendix F with permission of RIBA Publications Ltd, Finsbury Mission, Moreland St, London EC1V 8UB. The NJCC Code of Procedure for Single-stage Selective Tendering is reproduced in Appendix G with permission of the National Joint Consultative Committee for Building, 18 Mansfield Street, London W1M 9FG.

1

The contract documents

A landscape architect designs a scheme and arranges for a landscape contractor to contract to carry it out. The work to be executed under the contract is that stated in the contract documents. If it is in the contract documents it will be carried out; if it is not included it will not be carried out unless the contractor is paid extra for it. If it is not clear what is in the contract, interminable arguments will ensue. More important, the goodwill and trust between the parties (with the consequent loss of quality) may vanish overnight.

On the other hand if the contract documents are crystal clear the project has every chance of success and will proceed as smoothly as on well-oiled wheels.

The contract documents comprise:

1. A **specification** giving specific details as to the exact quality of workmanship and materials to be provided – *what*.
   Schedules of Quantity listing in similar detail the quantity of material to be provided – *how much*.
2. **Drawings** showing precisely the location of the materials – *where*. These are brought together by:
3. The **Articles of Agreement** and **Conditions of Contract,** setting out the obligations of the parties, when it is to be carried out, how it is to be controlled, paid for, insured against accidents during its execution and what happens in the event of a dispute between the parties.

1.1 Common Arrangement (CCPI)

In 1979 a committee was set up at the instigation of the NCC Standing Committee to co-ordinate project information. It in turn established as a first priority the need to devise a common arrangement of up to 200 work sections grouped into about 60 categories. These in turn could be brought together under 10–15 broad headings similar to the present trade headings. The intention was to ensure that drawings, specifications, and Bills of Quantities were related to each other with a common structure. In the event this resulted in Group D (Ground work) with 5 subgroups and Group Q (Site surfacing/Planting/Fencing) with 4 subgroups.

Those primarily affecting the landscape industry are D20 (Excavating and soil/hard filling) with 20 subsections, Q30 (Seeding/Turfing) with 15 subsections, and Q31 (Planting) with 20 subsections. These are listed in Tables 1.1–1.3, and provided that those items in each section relevant to a particular project are covered (in the same sequence) in drawings, specifications and Bills of Quantity, estimators will then be able to submit tenders confident in the knowledge that everything required of them will have been priced, and site staff know the location and quality of materials and workmanship required of them. Tables 1.1–1.3 are reproduced with permission of the Co-ordinating Committee for Project Information. The information in the tables is taken from the Common Arrangement, published in 1987.

Table 1.1 Common Arrangement: Subgroup D20 (Excavating and soil/hard filling)

Forming bulk, pit, trench and surface area excavations other than for m and e services. Filling holes and excavations other than those for m and e services. Making up levels by bulk filling or in layers, including hardcore but excluding bases to roads and pavings.

Included	Excluded
Applying herbicides to soil before excavating. Excavating topsoil, subsoil, made ground or rock.	Excavating and filling for temporary roads and other temporary works. (Preliminaries)
Breaking out existing substructures.	Ground investigation, D10.
Breaking out existing hard pavings.	Soil stabilization, D11.
Removing existing underground storage tanks and services.	Stabilizing soil in situ by incorporating cement with a rotovator. (Soil stabilization, D11)
Removing existing trees, shrubs, undergrowth and turf.	Site dewatering required to be executed by a specialist firm. (Site dewatering, D12).
Consolidating bottoms of excavations.	Excavating, backfilling, beds and surrounds for engineering services, D21.
Trimming excavations to accurate shape and dimensions.	Excavating and backfilling for underpinning. (Underpinning, D40)
Keeping the excavations free of water (other than site dewatering).	Filling internal plant containers. (Interior planting, N14)
Temporary diversion of waterways and drains.	Excavating and backfilling for below ground drainage. (Drainage below ground, R12)
Benching sloping ground to receive filling.	Excavating and backfilling for land drainage. (Land drainage, R13)
Making stockpiles of excavated material.	Hardcore/Granular bases to
Disposing of surplus excavated material.	
Earthwork support.	
Filling with and compacting:	
Excavated material (including selected);	
Imported material;	

Rock;
Hardcore and granular material;
Soil cement.
Consolidating surface of filling with fine material.
Preparing subsoil by ripping, grading, etc. before spreading topsoil.
Transporting from stockpiles or importing and spreading topsoil.
Filling external planters, beds, roof gardens with soil and drainage layers, including filter mats.

roads/pavings, Q10.
Gravel/Hoggin roads/pavings, Q22.
Cultivating and final fine grading of soil for seeding, turfing or planting.
(Relevant sections, Q30, Q31)
Spreading peat, compost, mulch, fertiliser, soil ameliorants, including working in.
(Relevant sections, Q30, Q31)
Excavating and filling treepits.
(Planting, Q31)
Excavating and filling post holes for fencing.
(Fencing, Q40)

Table 1.2 Common Arrangement: Subgroup Q30 (Seeding/Turfing)

Preparing soil and seeding or turfing to form lawns and general grassed areas.

Included

Applying herbicides.
Cultivating topsoil, including removing large stones and weeds.
Fine grading topsoil.
Providing, spreading and working in peat, manure, compost, mulch, fertiliser, soil ameliorants, sand, etc.
Edging strips for lawn areas.
Mesh reinforcement for grass areas.
Seeding and rolling.
Turfing, including turf edges to seeded areas.
Hydro-seeding.
Grass cutting.
Work to existing grassed areas, including scarifying, forking, fertilising, applying selective weedkillers, local re-seeding or re-turfing, rolling, edging, etc.
Protecting new and existing grassed and planted areas with temporary fencing, etc.
Watering, including during Defects Liability Period.
Work specified to be executed during the Defects Liability Period.
Replacement seeding and turfing.

Excluded

Earthworks and other preparation:
Clearing existing vegetation;
Removing existing hard paving and obstructions;
Removing existing topsoil and stockpiling;
General site contouring and adjusting levels;
Transporting from stockpiles or importing and spreading topsoil;
Filling external planters, beds, roof gardens with soil and drainage layers.
(Excavation and soil/hard filling, D20)
Land drainage, R13
Irrigation, S14
Grass block paving:
Sub-base
(Hardcore/Granular bases to roads/ pavings, Q10);
Blocks and bedding
(Slab/Brick/Block etc. pavings, Q24).
Special surfacings/pavings for sport, Q25.
Maintenance (as distinct from making good defects) other than that specified to be carried out during the Defects Liability Period.

Table 1.3 Common Arrangement: Subgroup Q31 (Planting)

Preparing soil and planting herbaceous and other plants, trees and shrubs.

Included

Apply herbicides
Cultivating topsoil including removing

Excluded

Earthworks and other preparation:
Clearing existing vegetation;

Table 1.3 (*Contd*)

large stones and weeds. Fine grading topsoil. Forming raised or sunken beds, borders, etc. Providing and spreading peat, manure, compost, mulch, fertiliser, soil ameliorants, and working in if required. Planting herbaceous plants and bulbs. Planting shrubs and hedges. Planting nursery stock and semi-mature trees. Excavating and backfilling tree pits. Fence supports for hedges. Support wires, etc. for climbers. Tree stakes, tree guards, tree guys. Wrapping and other protection of trees, shrubs and plants. Labelling. Applying anti-desiccants. Tree surgery, thinning and pruning. Protecting new and existing grassed and planted areas with temporary fencing, etc. Watering, including during Defects Liability Period Work specified to be carried out during the Defects Liability Period. Replacement planting.	Removing existing hard paving and obstructions; Removing existing topsoil and stockpiling; General site contouring and adjusting levels; Transporting from stockpiles or importing and spreading topsoil; Filling external planters, beds, roof gardens with soil and drainage layers. (Excavation and soil/hard filling, D20) Interior planting, N14 External prefabricated plant containers (Site/Street furniture/equipment N30). Land drainage, R13. Irrigation, S14 Tree grilles. (Slab/Brick/Block/Pavior/Sett/Cobble pavings, Z24 Seeding/Turfing, Q30. Fencing, Q40. Maintenance (as distinct from making good defects) other than that specified to be carried out during the Defects Liability Period.

1.2 The types of contract

Unfortunately of the three criteria of any landscape contract, **Cost**, **Quality** and **Time**, all three can rarely be achieved in any one contract at the same time and it is essential to ascertain at the outset which two the client requires to be given priority at the expense of the third. Certainty of cost and quality can be achieved through the use of **lump sum** forms of contract but it takes longer and loses the certainty of achieving early completion. Quality and early completion can be achieved by the use of **prime cost** or **management** forms of contract but with loss of the certainty at the outset of the project of the final cost. Early completion at the minimum cost can be achieved by the use of **design and build** contracts at the possible loss of quality due to the client's loss of control over the choice of landscape architect, other consultants, subcontractors and suppliers.

It is however fortunate that on the majority of landscape projects there is usually sufficient time to prepare comprehensive and accurate drawings and specifications, with the time for the execution of the project set by the seasons of the year and not the impatience of the client or main contractor. As a consequence most landscape contracts are let on a lump sum basis and it is this type of contract which is covered in detail in the following chapters.

There has been since 1978 a new (and hopefully more concise and more comprehensible) Standard Form of Contract for Landscape Works. It is called the **Standard Form** – let us all keep it that way. This is a contract agreed between the landscape architect and the contracting side of the industry, the clauses of which are a compromise attempting to strike a fair balance between both parties. It attempts to include only clauses which are universally applicable and acceptable on all contracts and any alterations therefore will almost certainly mean increased tenders.

This is not to say that experience of its application and changing circumstances, practices, procedures and economic forces on the industry will not involve amendments, e.g. vandalism (if it continues to increase at its present rate). These amendments must and should happen. They are, except in the most extreme cases, made on a 12-monthly basis (and only then after discussion and agreement between all sides of the industry).

It must also be realized that there may be some instances where a particular aspect of a contract justifies an alteration to the standard clauses; but before doing so, careful consideration must be given to the implications: the almost certainly increased cost and possibility of rejection either by one or all of the contractors, or by a local authority or government department on the part of the clients.

1.3 The contract drawings

It should not be forgotten that working or production drawings indicate to the contractor the exact location of the materials included in the specification; they are not the drawings prepared to show the client what it will look like when it is finished (or, as so often happens, five years after it is finished).

It must also be remembered that if there is written information on the drawing it must be clear, legible and intelligible. The chances are that it may be referred to, on site in the depth of winter – just before dusk and if it can not stand up to this test it is not worth the dyeline paper it is to be printed on. Either it will not be referred to at all or it will be referred to and then misread – there is no other alternative, and both are equally disastrous.

If instructions are not to be written but drawn, it is essential to use the same shorthand of graphic symbols as BS 1192: Part 4 (Recommendations for landscape drawings) otherwise there is as little chance of communicating one's intentions as if the foreman were being addressed in Esperanto. Novel coding systems should be resisted also, there is a very good one already: it is called the alphabet, whose symbols are combined to form things called words.

Always use figured setting-out dimensions and centres – the foreman has no adjustable set-square or compass on site!

1.4 The contract specification

Just as accurate drawings indicating the location of all materials and plants are essential for the successful outcome of a project so is a specification describing the minimum acceptable quality of materials and workmanship required to be provided by the landscape contracted, priced and included in his tender. Any ambiguities, errors or omissions on the part of the landscape architect at this stage will ensure the successful outcome of his client's project is probably doomed from the start.

Fortunately the selection of standard clauses appropriate to each particular project from a library of landscape standard specification clauses such as those produced by NBS or PSA in the Common Arrangement, coordinating drawings specification and Bills of Quantities (if required) are followed and the selection is conscientiously carried out, this is unlikely to happen.

2

Landscape contract drawings

Of the four components of a set of contract documents: the illustrative, the qualitative, the quantitative and the conditions (or as they are more comprehensibly known: the drawings, the specification, the Bills of Quantity, and the Form of Contract), the drawings are the one aspect on which the landscape architect considers he needs little or no guidance, even if this view is not so universally held by landscape contractors.

Neither will however deny the statement in the foreword to BS 1192: Part 4: 1984 *Landscape Drawing Practice* – even if they were previously unaware of its existence – that the recommendations regarding the production of graphical information are needed:

> . . . to provide communication with accuracy, clarity, economy and consistency of presentation between all concerned with the construction industry, including architects, civil engineers, contractors, landscape architects, services engineers, site operatives and structural engineers . . . It is hoped that this Part of BS 1192 will encourage the standardisation of drawings and schedules. Familiarity with commonly accepted forms of presentation is likely to result in greater efficiency with the preparation and interpretation of landscape drawings and minimise the risk of confusion on site.

The translation of a landscape architect's design into reality by the landscape contractor is like painting a portrait by numbers with the contractor holding the brush, and the designer trying to tell him in words what colours to use and where to put them. This is similar to instructing an orchestra on the sounds they are to make when playing a symphony through the accepted convention of a musical score no more than a range of dots of various shapes and sizes arranged in accordance with a universally accepted convention across a series of parallel lines on a piece of paper. It is a matter of some regret that the same effort is not devoted to learning an equally universal set of conventions acceptable to the landscape industry.

2.1 Graphic symbols

A commonly accepted set of conventions to indicate on landscape drawings, by means of symbols and abbreviations, the most commonly used elements of landscape work (both hard and soft) is essential to the landscape industry. Their universal adoption on production drawings will convey the maximum information in the minimum time and effort, clearly and with the minimum risk of misunderstanding and error. BS 1192 Table 2 sets out the 50 symbols most commonly found to be needed (Figures 2.1 and 2.2). Irrespective of personal preference the temptation to improve certain of the more obscure must be strenuously resisted if all landscape contractors throughout the British Isles are to recognize instantly the work they are required to do on all projects, irrespective of the different offices from which the drawings have originated.

For new planting, the choice of symbols is dependent on the range and complexity of the work and the scale at which it has been drawn. Currently the reproduction of drawings still has to be in one colour and identification has therefore to be by symbol, hatching or by abbreviation. Colouring the multiple copies that are required on a landscape contract is costly, time-consuming and the risk of error rises in direct ratio to the number of copies required. Rarely should colour therefore be used on any landscape drawings other than those which are required for presentation purposes.

At whatever scale the drawing has been prepared (and one must not overlook the ability to photo-reduce or enlarge drawings on many commercial photocopiers) the lines, symbols, dimensions, figures and text must always be capable of reproduction clearly and legibly.

(Note detailed guidance is also given in BS 1192: Part 4 on schedules (see pp. 32–33), reference grids, interpolation of contours, visibility between given points and drawing contours in perspective.

2.2 Scope

In contrast to BS 1192, which covers all aspects of landscape drawings, it is only intended in this chapter to cover those drawings which are used for the landscape contract itself although the same principles will of course apply to those used by the architect and landscape architect for his discussions with his client and with local and other authorities in the exercise of their statutory obligations.

2.3 Size and content

Production drawings for landscape contracts should be prepared on a standardized sheet chosen, irrespective of the size of the site or project, for ease of handling on site compatible with the need to present as much information on as

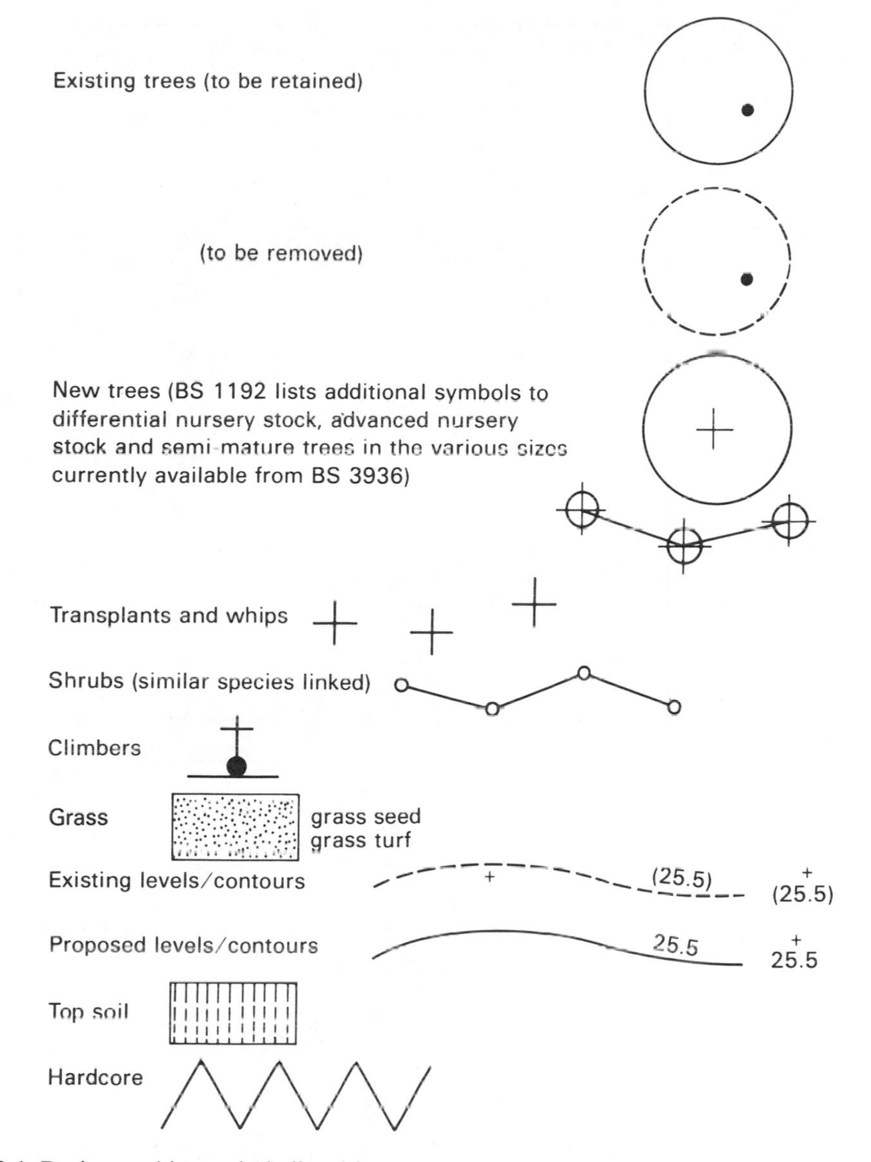

Fig. 2.1 Basic graphic symbols listed in BS 1192 : Part 4. (Note that BS 1192 provides in addition to the basic symbols indicated here for the drawings to show also, by the additional details included in Figure 2.2, the differentiation between transplants, nursery stock and semi mature trees and even the sizes within each group. It remains to be seen if, in view of their complexity and the possibility of error, these will be widely adopted.)

few sheets as possible. For most projects A1 and A2 sizes for location, site, setting-out and layout plans, with A3 and A4 for assembly component drawings and planting plans, will be appropriate at scales from 1:50 000 down to 1:200 for the former, and between 1:200 and 1:5 for the latter.

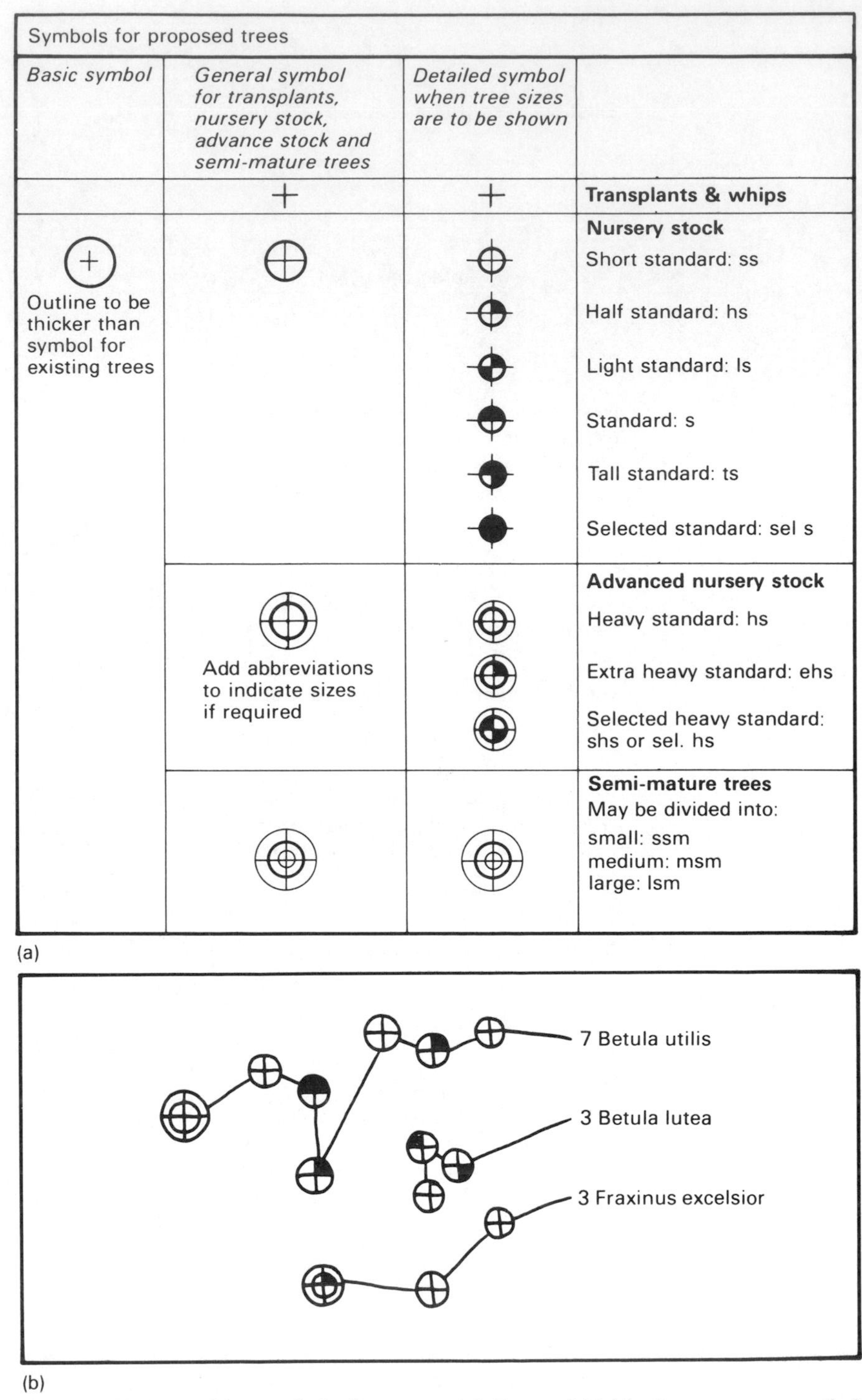

Symbols for proposed trees			
Basic symbol	*General symbol for transplants, nursery stock, advance stock and semi-mature trees*	*Detailed symbol when tree sizes are to be shown*	
	+	+	**Transplants & whips**
Outline to be thicker than symbol for existing trees			**Nursery stock** Short standard: ss Half standard: hs Light standard: ls Standard: s Tall standard: ts Selected standard: sel s
	Add abbreviations to indicate sizes if required		**Advanced nursery stock** Heavy standard: hs Extra heavy standard: ehs Selected heavy standard: shs or sel. hs
			Semi-mature trees May be divided into: small: ssm medium: msm large: lsm

(a)

(b)

Fig. 2.2 Detailed graphic symbols for trees. (a) Part of Table 2 on recommended tree symbols, reproduced from BS 1192 : Part 4.

Production drawings must be concise and have been prepared so that they remain legible at all times of the day (and night) and at all seasons of the year, both in the contractors' hut and out on site, for that is where they are intended to be used. They must contain sufficient information for the design to be implemented and for accurate quantities to be taken off and measured by the quantity surveyor and/or the estimator.

No rules exist at present regarding the content of drawings, although work has started under the aegis of the Coordinating Committee for Project Information (CCPI) which has attempted to synthesize the rules for drawings, specification and method of measurement under a Common Arrangement. BS 1194: Part 4 however includes check lists (Table 2.1) which attempt to itemize the minimum information that should appear on the various types of drawings.

Drawings should only in exceptional circumstances duplicate information already indicated elsewhere – on other drawings and in the specification or schedules – in view of the risk of error in the event of subsequent changes being made on one document and not on the other (irrespective of the duplication of effort needed to make the same change twice). Even the temptation to repeat the same information to more detail on a subsequent drawing to a larger scale – with the requirement in the Bill or specification: 'Always work to the largest scale drawings' – is to be avoided in view of the contractors' undeniable preference (irresistible temptation?) to use the most comprehensive drawing with the maximum information on one sheet (i.e. the one to the smaller scale).

Table 2.1 Checklists of information required on various types of drawings

Type of drawing	Information required	Preferred scale
Location plans:	Site location, access and site surroundings	–
Site plans:	Existing and proposed building, fencing and hedges within and adjoining the site, paths, roads, trees, shrubs, grass and water features	1:500
Site work plans:	Contractors' working area, existing trees, other features to be maintained and protected. Site restrictions as to location of temporary accommodation, surplus excavated material and top soil	1:200
Setting-out plans:	Reference points levels and setting-out dimensions for hard and soft landscape	1:200
Layout plans		
I Hard landscape:	Hard surfaces, walls, fencing, gates, curbs and edging, location of details	–
II Soft landscape	New and existing levels and contours, depths of topsoil, irrigation and land drains, tree, shrub, herbaceous, grass areas and water features	–
III Services:	New and existing soil and surface water drainage, gas, water, electricity and telephone services	1:100/1:50
Assembly and construction details:	Plans, sections and elevations enlarged as necessary to show details of construction and assembly	1:20/1:5
Planting plans and component details:	Location, number, size and species of trees hedges, shrubs and herbaceous plants, grass seeded and turfed areas	1:20/1:5

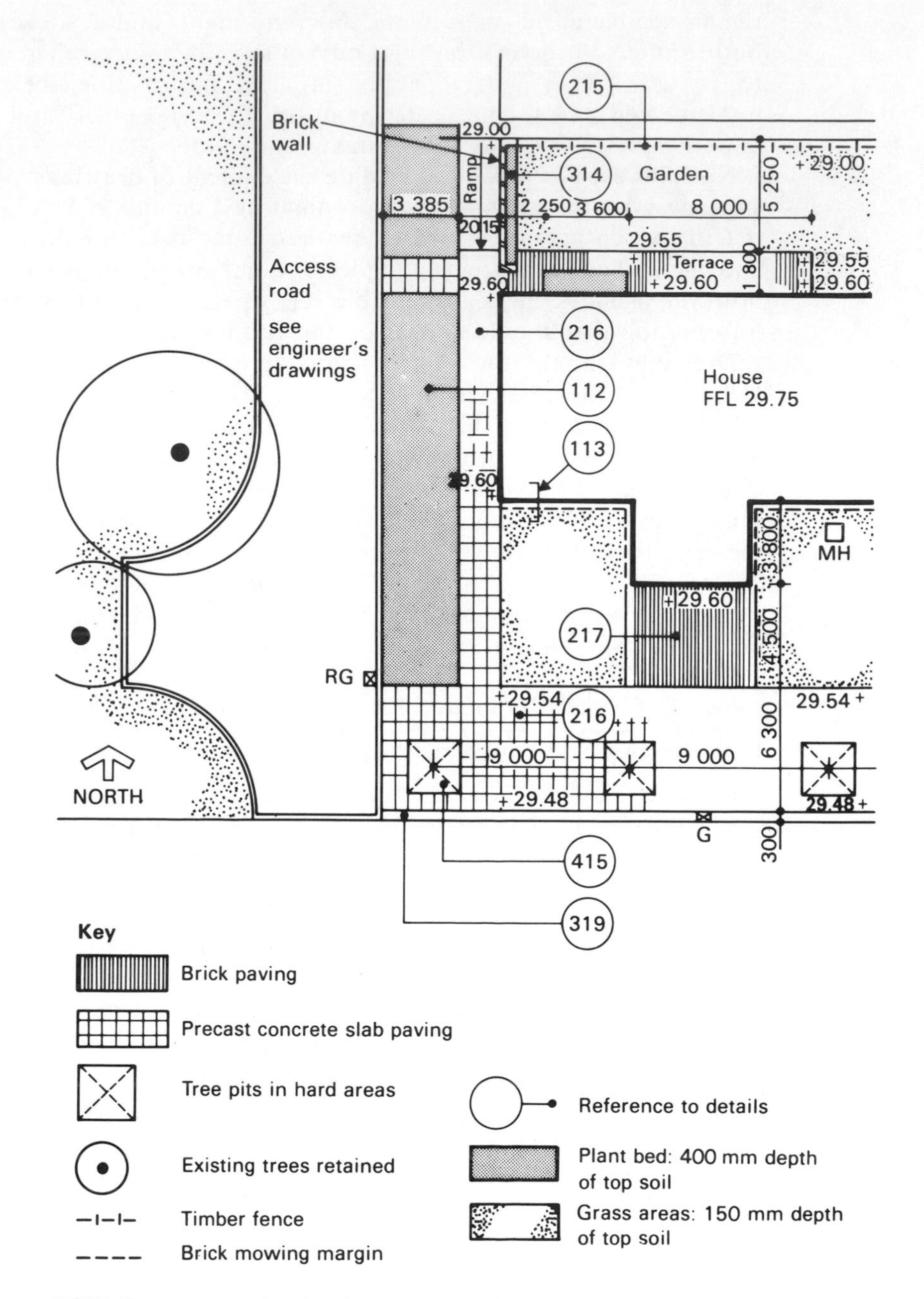

Fig. 2.3 Example of hard-landscape plan. (After BS 1192 : Part 4 : 1984.)

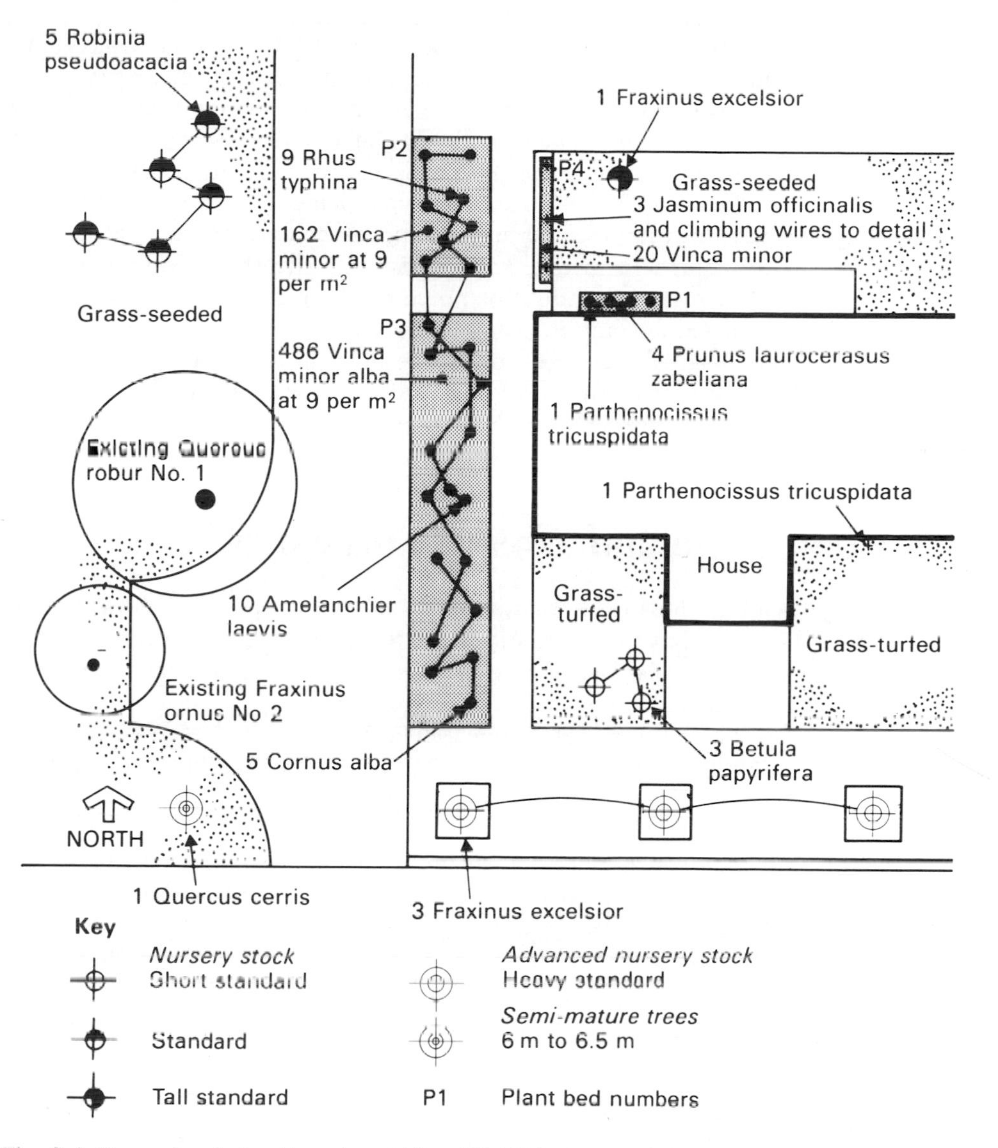

Fig. 2.4 Example of planting plan. (After BS 1192 : Part 4 : 1984.)

2.4 Layout Drawings

Layout drawings must always be crystal clear and in some instances it may be helpful if successive elements of the work are shown in separate sheets, e.g. site clearance, setting out, hard landscape, planting, etc. (Figures 2.3 and 2.4 show examples.) The use of copy negatives, and overlay draughting techniques, with one basic drawing from which copies are made before adding the additional note is particularly useful in this connection.

Discrepancies on drawings indicating the extent of cut-and-fill frequently give rise to problems on landscape contracts and it is essential that this is accurately indicated both on plan and section with the utmost clarity if such disputes are to be avoided (Figure 2.5). The location of the details for which additional component and assembly details have had to be prepared should be indicated.

Most landscape projects will involve the treatment of both the horizontal surfaces and the vertical enclosing elements of walls and fencing. The latter will inevitably involve the preparation of the larger scale component details which are best prepared in the form of separate sheets of component details (Figure 2.6). Their existence should be noted on the relevant locations on the layout drawings by a section line in conjunction with the detail reference number enclosed in a circle. It is of course essential to indicate on the layout drawings the exact extent of the surface finishes both hard and soft, and the precise length of all enclosing walls and fences.

2.5 Component and assembly drawings

While these often may be a set of photocopies of relevant standard details with or without the name of the particular project inserted in the job title panel, they may also be issued having been traced onto a larger sheet for inclusion in the contract documents. In either case there is no doubt of their relevance to any particular contract, whereas their inclusion by reference to a standard library of details, both relevant and inapplicable, available for inspection on demand will always cause problems particularly if they are amended subsequent to the date of the contract.

It is essential that the contractor has all the drawings at the time of tender preparation for any omission can lead to serious delays in progress and a claim situation. All drawings must be carefully numbered or referenced and these identifications noted in the specification and Bill of Quantities.

2.6 Curves

Curves that are drawn roughly freehand on preliminary drawings are never satisfactory when the time comes for them to be translated on site if accurate dimensions for their setting out are not provided by the designer. 'French curves', which are so easy to use on the drawing board, do not exist on the ground. All curves, whether derived from circles with compasses or templates (circular curves) or irregular (non-circular) curves that must rely on dimensioned offsets, therefore must always have their origins clearly indicated on the drawings with radii, tangent points for circular curves and closely spaced offsets for irregular curves clearly dimensioned (Figure 2.7).

In practice on most landscape projects, carefully selected circular curves will be found to suffice in the majority of instances, irregular curves set out by offsets

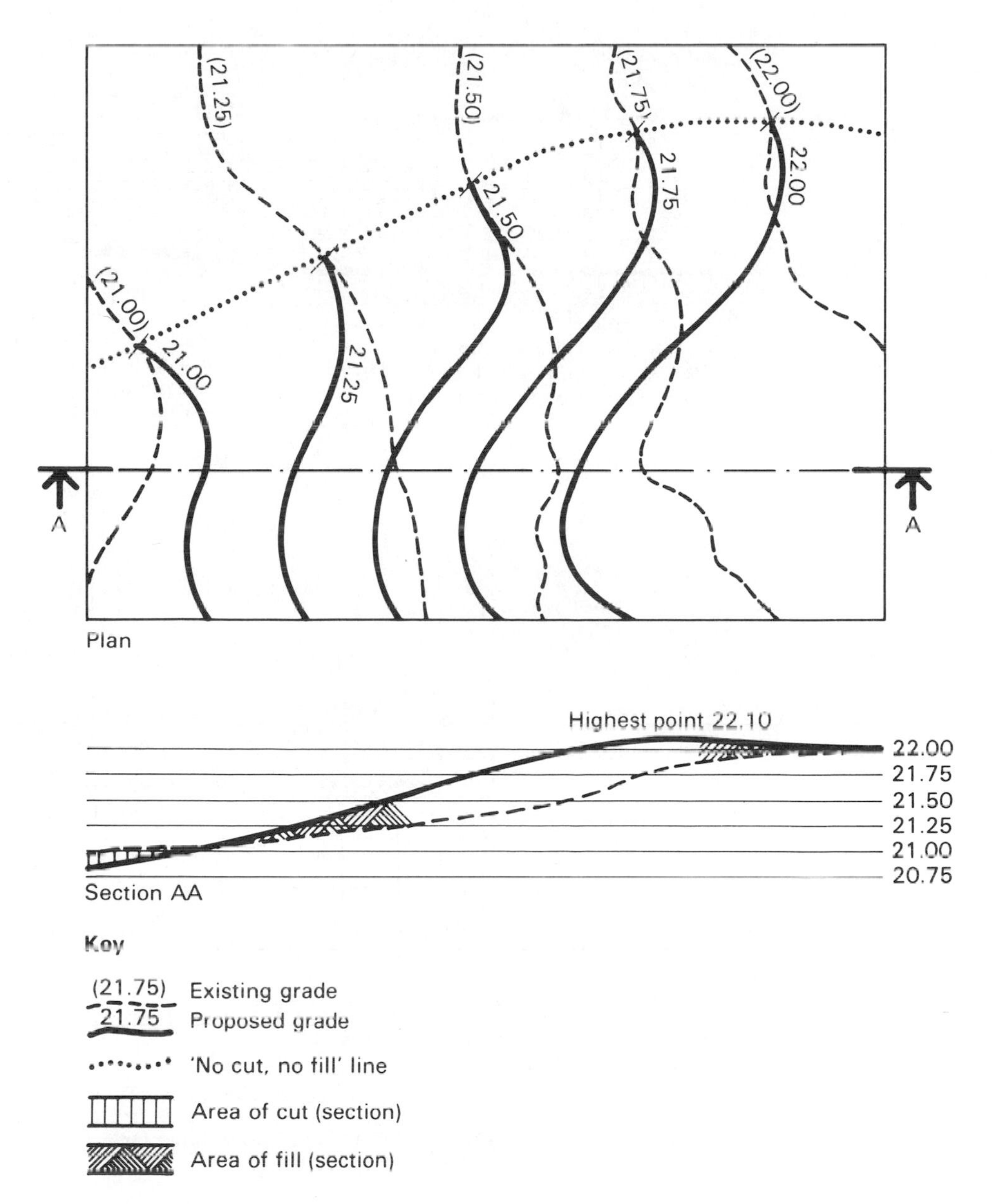

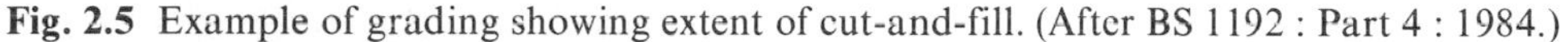

Fig. 2.5 Example of grading showing extent of cut-and-fill. (After BS 1192 : Part 4 : 1984.)

(Figure 2.8) or coordinates of points along the curve being required only for the larger projects involving major recontouring or major road layouts.

Irrespective of whatever type of curve is used the sizes and locations should be clearly indicated and dimensioned, without the need to scale off the drawings – a method which is never sufficiently accurate and inevitably results in much 'trial and error' on site, expensively wasting the time of all concerned.

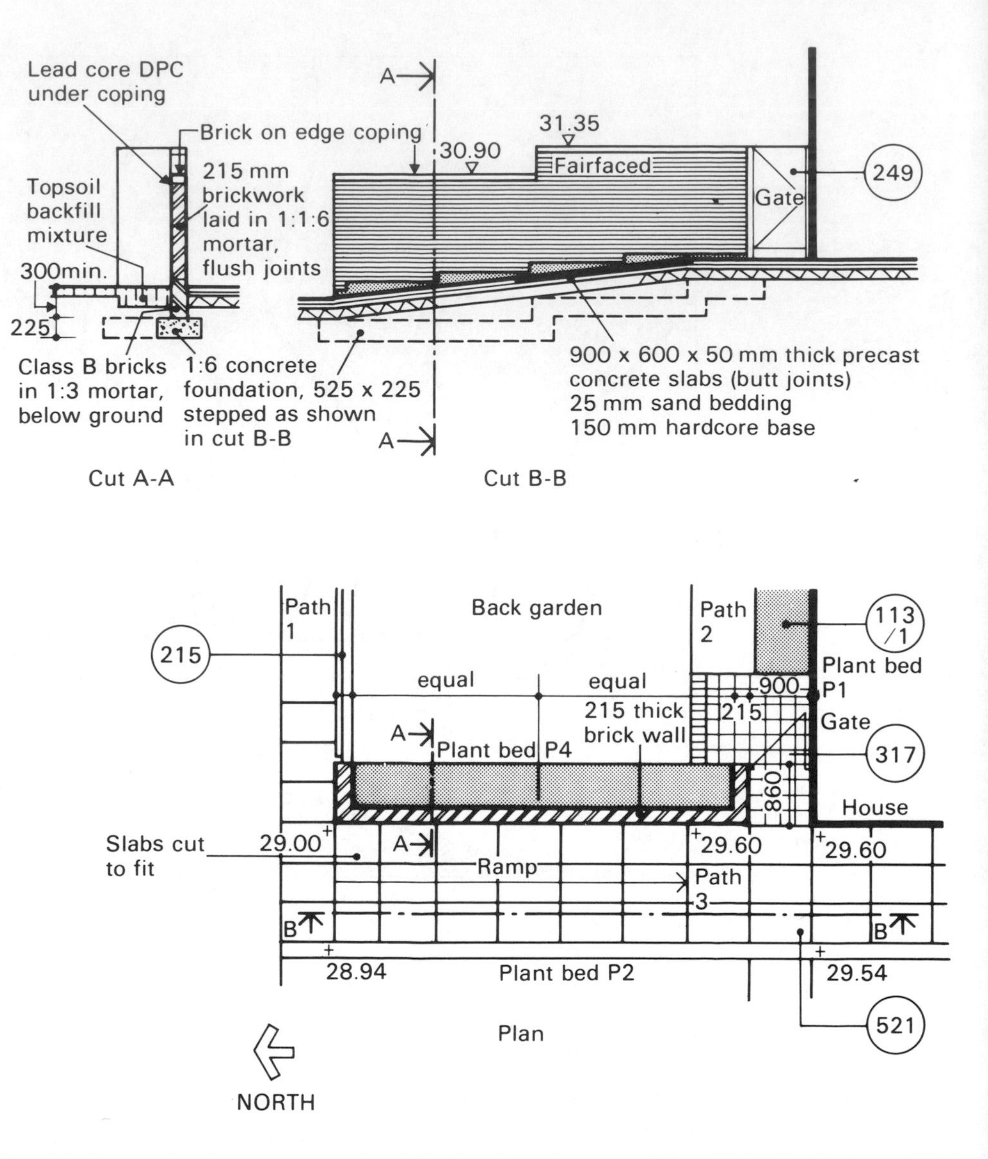

Fig. 2.6 Example of component detail for hard landscape. (After BS 1192 : Part 4 : 1984.)

2.7 Setting out

Accurate setting out is crucial to the success of every project and a clearly dimensioned drawing must always be the first priority. Where existing features on site are involved, or setting out relies on certain aspects being set out or aligned by eye this must also be clearly indicated (Figure 2.9) so that those responsible for setting out clearly understand the intentions of the designer.

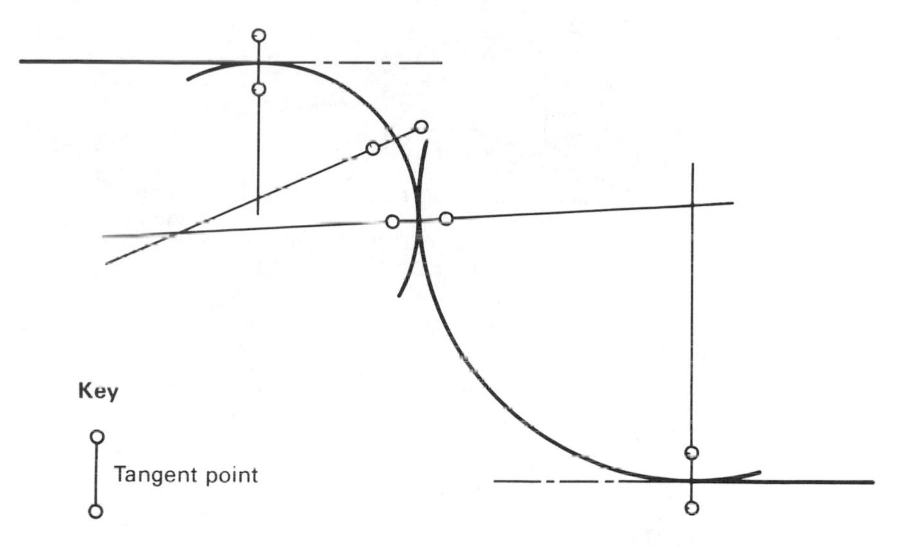

Fig. 2.7 Drawing of circular curves, showing tangent point. (After BS 1192 : Part 4 : 1984.)

2.8 Levels

The indication of new and existing levels must be a key feature of any landscape project. Whereas the addition of contours to a plan gives a clear indication of the designer's intentions, spot levels at precisely identifiable locations are usually equally essential for those responsible for the execution of the work on site. The conventions that new levels should be shown unenclosed, e.g. 32.1, existing levels being enclosed in brackets or boxes, e.g. (32.1), new contours being shown by a

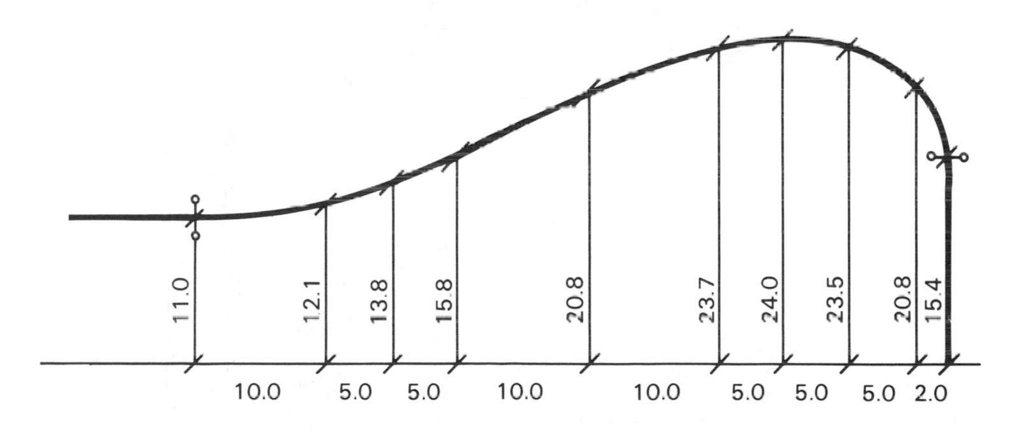

Fig. 2.8 Location of an irregular curve by offsets. (After BS 1192 : Part 4 : 1984.)

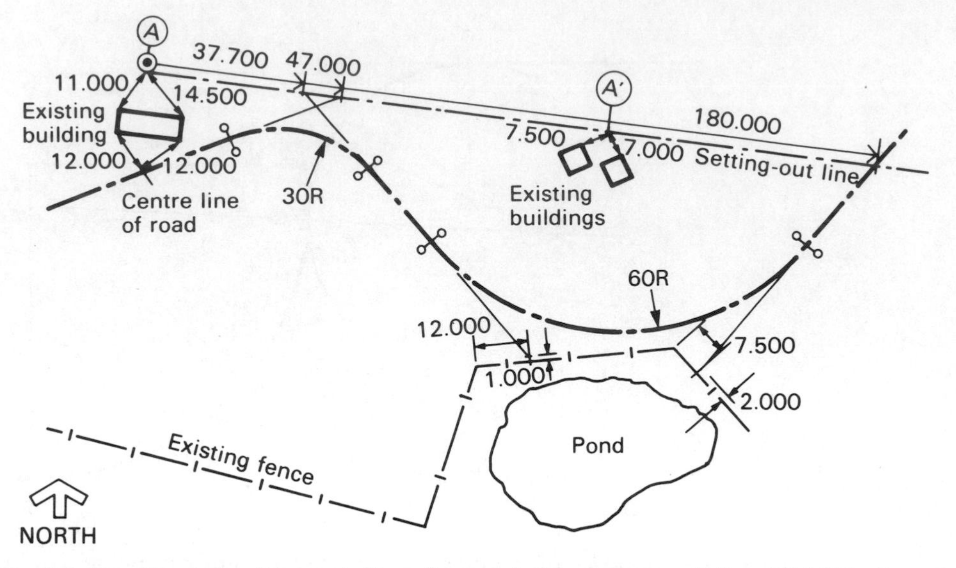

Fig. 2.9 Location of setting-out lines from existing features. (After BS 1192 : Part 4 : 1984.)

solid line, and existing contours dotted. Arrows pointing 'up' the slope must always be followed if mistakes are to be avoided (see Figures 2.5 and 2.6 for examples).

3

Landscape contract specification

The word **specification** is defined in the Concise Oxford Dictionary* as:

specification – . . . specified detail, esp. (in *pl.*) detailed description of construction, workmanship, materials, etc., of work (to be) undertaken, prepared by architect . . ..

Clearly, therefore, when used as a contract document, the specification must describe precisely and unambiguously the quality of materials, workmanship and in some cases the performance of a particular element or component required to be provided. The specification must always be clearly identifiable so that the purpose and the needs of the specifier are immediately comprehensible; the contents must be set out in an organized and orderly way so that the requirements of the specification are immediately understood and priced accordingly by the estimator.

The description of the quality of the workmanship and materials required is as essential a part of the contract documents as the drawings and Schedules of Quantity in conveying the designer's intentions to the contractor, and to ensure that he has allowed in his tender for the proper execution of each element of the work.

It must be obvious therefore that the only member of the team able to undertake the proper preparation of the specification element of the contract is the designer and only rarely can he delegate this aspect of his work to another, such as the quantity surveyor.

It is unfortunate that in some earlier Forms of Contract where Bills of Quantities were included in the contract documents the 'specification' was expressly excluded, the reason being that in those instances the specification of the quality of the work had been included elsewhere, i.e. in the trade preambles to the measured items prepared by the quantity surveyors. Fortunately this

*Concise Oxford Dictionary, 5th edn (1971) Oxford University Press, Oxford

anomaly has now been removed and all Standard Forms of Contract now provide for the specification of workmanship and materials to be included either as trade preambles or as a separately identified contract document whether Bills of Quantities are prepared by the employer or not. In either case it is important to ensure that all elements of qualitative description are kept clearly separate from the measured items, even if this seems illogical – as in the instance of the measurement of facing brickwork where the tenderer is referred to the specification clauses for the type or make of brick to be included. Unless this rule is strictly observed, however, as soon as some aspects of quality appear in the measured items, the temptation becomes irresistible on the part of the contractor to overlook, forget or ignore the specification in the day-to-day pursuance of his work.

3.1 Clause selection

The essence of a good specification is achieving the right balance as regards both quantity and quality and the right size for the job.

Its preparation requires great skill and judgement on the part of the draughtsman. From a vast library of clauses he has to select those applicable to that particular project (nothing shows up lack of experience and lack of confidence more quickly than a specification incorporating clauses which are inapplicable or incapable of achievement). He has to be cautious of clauses specifying in too much detail some comparatively unimportant item, e.g. four pages on tree surgery when only one limb of one tree is involved. Secondly it is essential only to specify what one is competent to check one is getting and resist the temptation to pad out the specifications with clauses included as insurances to hide one's own ignorance; contractors will spot them a mile off and turn it to their own advantage!

If the contractor is to give a job the attention it deserves – give an equal amount of attention to the specification. Running off a few standard sheets in the hope that they will do, because one is too lazy, too inexperienced or too busy to have typed out a specification omitting the inapplicable clauses and (more importantly) adding those clauses which were not necessary on the last job but are for this. Always go back to the full library, not an abridged version of a specification of a previous project.

Equally, the Schedules must be practicable (do not schedule a 2.0-metre E.L.N.S. Ginko, Ilex or Taxus!) and accurate so that the tenders when submitted are comparable.

3.2 Structure and phraseology

It is now a generally accepted convention that specifications are divided into two. The first part comprises the 'preliminary' clauses covering the Form of Contract to be used, the responsibilities of the parties to the contract, the definition of the procedures to be followed as a consequence, and a simple description of the scope of the work, the site and any particular restrictions as to access, sequence of the work, hours or working rules. The second part of the specification describes the quality and performance of the materials, workmanship and 'prefabricated' components to be used. These are best set out by describing the materials and components for any particular element followed by the workmanship required in the preparation of adjacent surfaces, the location, assembly, planting and subsequent protection of the element itself. Other aspects which may have to be described are those affecting appearance, performance and any requirements for off-site or on-site testing, tolerances, and the provision and retention of samples.

The clauses themselves must also be written in clear and concise phrases, and words chosen to indicate the designers' intentions simply and unambiguously. They should, wherever possible, start with a 'Keyword' indicating the scope of the clause, e.g. 'TOPSOIL', 'WEEDS', 'HERBICIDES', followed by a verb in the imperative, e.g. 'to be', 'provide', 'lay', 'fix', 'plant', etc. (Table 3.1).

Vague phrases such as 'in all respects', 'the best of their respective kind', 'free from all faults' should be avoided at all costs, lest they exceed the bounds of practicability or involve exorbitant cost far beyond that intended by the specifier. Specification clauses should never require anything not capable of being checked by the specifier ('policed') nor should meaningless phrases such as 'and the like' be used.

Words should always be used consistently, the same word being used for the same thing throughout the specification, and the same word or term used where it occurs elsewhere in the other contract documents and standard clauses.

Clauses should always be concise using the minimum number of words to convey their intended meaning without ambiguity, i.e. 'Arrange . . .' and not 'The contractor shall make all arrangements as are necessary to ensure . . .'. Describing operations in great detail or to an accuracy or tolerance that is unnecessary or impossible to attain should always be avoided.

3.3 Materials

Materials can be specified in one of three ways: by type, by British Standard, by proprietary name. Whichever is selected the minimum standard must be clear, but it is always in the employer's interest to allow a contractor the maximum freedom of choice in the selection of his sources of supply.

Table 3.1 Example of specification clauses for soiling/cultivating/grading (reproduced with permission of NBS Services Ltd)

This section covers:	To be read in conjunction with
• Preparation of subsoil • Spreading of topsoil Initial treatment, cultivation and grading of in situ and spread topsoil.	Preliminaries and General Conditions.

Table 3.1 (*Contd.*)

For site clearance, excavation and stockpiling of topsoil see section C12. For subsoil filling see section C21. For seeding and turfing see section W21. For planting see section W31. The soiling/cultivating/grading work may be carried out: • Entirely by the Main Contractor • Entirely by a landscape Nominated Subcontractor • Partly by the Main Contractor and partly by a landscape Nominated Subcontractor. In the latter case Section W11 can be used to produce separate specifications for the Main Contractor (e.g. preparing subsoil and spreading topsoil) and for the Nominated Subcontractor (e.g. cultivating and grading).	W11 :	**Products/Materials** *Y Soil/Additives/Herbicides*
Y210 Topsoil is the original surface layer of grassland or cultivated land. It does not generally include soil from woodland, heathland, moorland, bog or other special areas such as land impaired by industrial activity. BS 3882 classifies topsoil in relation to the following characteristics: • Texture: Insert 'light', 'medium' or 'heavy' • Soil reaction: pH value of neutral soil is 7.0 • Stone content: In assessing whether topsoil is reasonably stonefree, the natural local and site conditions should be taken into account • Size of stones: for ordinary purposes stones up to 50mm in size are acceptable.	Y210	Imported topsoil: to BS 3882. Texture: . . . Soil reaction: neutral. Reasonably free of stones. Maximum size of stones in any dimension: . . . mm. Free of weed seeds, roots of perennial weeds, sticks, subsoil and foreign matter. From an approved source.
	Y220	Imported topsoil: obtain approval of a sample load of not less than 5m^3. Retain for comparison with subsequent loads.
Y230 Use this clause when the Contractor is required to import topsoil from a particular source, stating all relevant details, e.g. location, access, whether or not it has to be stripped or lifted from a spoil heap, price agreed (if any).	Y230	Imported topsoil: obtain from . . .

Y410, Y420 Peat and sewage sludge are included in this section as additives to adjust the texture of topsoil when necessary.	Y410	Peat: to BS 4156.
	Y420	Sewage sludge: friable, no large lumps, dry and containing not less than 6% nitrogen, 3% phosphate and 5% potash.
	Y610	Lime: fine ground limestone containing not less than 50% of CaO equivalent.
Y810, Y820 Alternative clauses. The ACAS list is published under the title of 'Approved Products for Farmers and Growers' published every year by HMSO for the Ministry of Agriculture, Fisheries and Food.	Y810	Herbicide: a type recommended for the purpose in the current list of the Agricultural Chemicals Approval Scheme.
Y820 Use this clause to specify herbicides by proprietary reference as required. Herbicides should be selected from the current list of the Agricultural Chemicals Approval Scheme. The clause can also be used to restrict use of herbicides to certain types for specific applications e.g. • Adjacent to watercourses, lakes, ponds, etc • Adjacent to crops • Areas accessible to the public • Where existing flora and fauna are to be protected.	Y820	Herbicide: . . .

In the case of plants it is clearly insufficient to specify just the type, e.g. to specify *Fraxinus oxycarpa* 'Raywood' or *Hedera colchica* 'Variegata' without any indication of age, size or source of supply.

Equally to specify that the plant be in accordance with 'the relevant' British Standard (Table 3.2 and Appendix A) is inadequate as alternative sizes are almost invariably included within the one Standard. The reference to the actual requirement has to be made clear, e.g. advanced nursery stock to BS 5236 : 1975. Even this allows both a 'heavy standard' tree to be provided to a stem circumference (measured 1 m from the ground) of between 12 and 14 cm, to be between 3.50 and 4.25 m high with a 1.8 m clear stem, and 'extra heavy' with a stem circumference of 14–16 cm and a height of 4.25–6.00 m.

The use of a proprietary supplier, while always to be used with caution, is often the only way of ensuring plants of an adequate standard are delivered. Just as the health of a child cannot be measured by its height, so the delivery of healthy plants cannot be ensured solely by specifying their minimum size. While this practice removes the competitive element of pricing from the contractor (and its use should be avoided wherever possible) it does, at least, ensure the comparability of this part of the tender, with all plants supplied the same minimum quality.

Table 3.2 Fifteen British Standards considered essential in landscape work. Many are unaware of the existence of the 83 British Standards for gardening, horticulture and landscape work or that 49 of them (i.e. 60%) are available in summary form. Fifteen of these Standards are considered essential references for all those concerned in landscape work; they are listed here and summarized in Appendix A.

1. *Nursery Stock* (FAC/1)

BS 3936 : Parts 1–11 : Specification for trees and shrubs, etc.

Part 1. Trees and shrubs
Part 2. Roses
Part 3. Fruit
Part 4. Forest trees
Part 5. Poplars and tree willows
Part 6. *not used*
Part 7. Bedding plants
Part 8. *not used*
Part 9. Bulbs, corms and tubers
Part 10. Ground cover plants
Part 11. Culinary herbs in containers

2. *Operations* BDB/5

BS 3998 : 1966	Recommendations for tree work
BS 4043 : 1978	Recommendations for transplanting semi-mature trees
BS 4428 : 1968	Recommendations for general landscape operations
BS 5236 : 1975	Recommendations for cultivation and planting trees in the extra large nursery stock category
BS 5837 : 1980	Code of Practice for trees in relation to construction
BS 737091/94	Ground maintenance Part 1 (1991) – Establishing and managing; Part 2 (1994) – Hard areas (excl. sports surf.); Part 3 (1991) – Amenity of functional turf; Part 4 (1993) – Soft landscape; Part 5 – Water areas

3. *Materials* (FAC/2)

BS 3882 : 1994	Recommendations and classification for topsoil
BS 3969 : 1965(78)	Recommendations for turf for general landscape purposes
BS 4156 : 1967(79)	Peat
*BS 5551 : 1978/81/82	Fertilizers

4. *Terminology*

BS 881/589 : 1974	Nomenclature of commercial hardwoods and softwoods
BS 1192 : 1984	Part 4 Recommendations for landscape drawings
*BS 1831 : 1969	Recommended common names for pesticides
*BS 2468 : 1963	Glossary of terms relating to agricultural machinery
BS 3975 : 1966/69	Glossary for landscape work; Part 4 – Plant description; Part 5 – Horticultural practice

* Not in Construction Digest Handbook

When an acceptable description intelligible to all parties cannot be given without resorting to involved text or site inspection of plants in the nursery of the supplier, then it is either necessary to specify a single source suffixed by 'or equivalent' or 'or equal and approved', or by the more preferable alternative of listing, in the tender documents, three or four named suppliers from one or other of whom the Tenderer must obtain the material in question. When it is considered acceptable to add the suffix 'or equal and approved' to such a list, the proviso must be added that 'approval of the alternative source must be obtained in writing from the specifier at least seven days before the latest date for the submission of tenders'.

3.4 Workmanship

When specifying minimum standards of workmanship a choice of three ways apply: by finished effect, by British Standard Codes of Practice, and by method. By finished effect, with full responsibility for the method of achievement placed with the contractor, is clearly preferable but rarely practicable. 'Plant one semi-mature *Fagus sylvatica*' is clearly inadequate. Even the addition of 'in accordance with BS 4043 : 1966 – Recommendations for transplanting semi-mature trees' is rarely sufficient. Most specifiers will wish to rely on their own expertise and, while they may not specify the sequence and time of planting, they will almost certainly wish to specify the minimum dimensions and materials to be used, and the method of application in the construction of tree pits (even if, by strictly complying, the contractor may therefore be relieved of his responsibilities for their subsequent survival).

3.5 Protection

The protection of plants after they have been planted and before they become established always causes particular problems on landscape projects, not only in respect of watering during the Defects Liability Period but equally in the provision of temporary protection. Such clauses as 'provide such temporary fencing to new shrub borders and grass areas during the works as is necessary and remove on completion of the Defects Liability Period' is today no longer good enough. While such a phrase might just possibly (but probably would not) suffice for the protection of an existing mature tree during the progress of the work, when applied to watering and the protection of new work, it is woefully inadequate.

When faced with such a loosely worded clause many contractors will conclude that the specifier does not rate this work of much importance nor rank it very high in the list of his priorities, and is therefore unlikely to enforce its provision with any vigour, if at all. Under such circumstances the prudent estimator who seriously intends to submit the lowest tender will allow little, if anything, in his tender, in the hope that he will be required to provide little more than a nominal amount of work for this item if any at all. Conscientious contractors will of course allow the proper amount and may, as a consequence, never submit the lowest tender and rapidly go out of business. The unscrupulous contractor, unexpectedly having the conditions enforced, will suffer serious financial loss, perhaps even going out of business himself. When the client then turns to the conscientious contractor to remedy the work he may find he no longer exists either; the project is, as a consequence, a disaster for all concerned.

As pointed out in JCLI practice note 3, where temporary protection such as fencing is considered necessary, this must be clearly specified and quantified at tender stage. Ownership of temporary protection, even if it is to be left in position during the Defects Liability Period, normally remains the property of the contractor unless stated otherwise in the contract documents; it is often an advantage if they are specified to become the property of the employer after practical completion.

3.6 Standard specifications

It must by now be clear that there is no such thing, or there ought not to be, as a standard landscape specification. What all offices have, or should have, is a standard **library** of specification clauses from which the appropriate clauses can be selected for any particular project.

The use of standard clauses throughout the landscape industry obviously has the advantage of ensuring familiarity both on the part of the specifier and on the part of the contractor or supplier. Each knows from past experience the responsibilities and requirements of the clause, without the need for the careful scrutiny a new clause needs to ensure that the full obligations of the words used are fully understood and appreciated. This applies both to clauses in project specifications and to those used in trade preambles to measured items in Bills of Quantity. The Property Services Agency (PSA), and the National Building Specification (NBS, Table 3.1) both have libraries of standard specification clauses prepared specifically for landscape work. Grouped into sections, they include every alternative that is likely to be necessary on the full range of projects. They have comprehensive sets of preliminary clauses, with notes adjacent to each clause giving guidance as to the selection of the most appropriate alternative in any given situation. The various British Standards, Codes of Practice, and sources of other technical information are listed to assist in the selection of the appropriate clause. They all classify and codify clauses in the Standard Method of Measurement (SMM) Common Arrangement sequence and subdivide them into the two subsections referred to earlier: namely the specification of the materials to be used in each section, followed by the workmanship required in the various operations necessary for its location, assembly or planting. Those provided by NBS have the advantage of being appropriate to all landscape projects in both public and private sectors, and have the added advantage of having been edited by members of the LI, whereas those of PSA are copyright and intended only for PSA projects. In addition NBS is available on subscription at a special rate to landscape offices which are not part of a larger architectural practice. All are regularly updated and revised to ensure that only the latest British Standard is referred to, and, where options exist within the clauses themselves, these are identified to enable the appropriate choice to be made.

It is therefore obvious that there should be no such thing as a 'standard office landscape specification' as distinct from an office library of standard clauses from which those appropriate are extracted for each particular project. This will ensure that all aspects of the project are adequately specified, which rarely happens when the specification from a similar previous job is used, deleting (hopefully) those clauses that are inappropriate for the current job and neglecting to add those that are necessary for the new project but were not required on the previous one.

There is however one danger that arises from the selection of clauses from a fully comprehensive library: that of overspecifying. There is a great risk, particularly when inexperienced or unconfident, to include clauses 'just in case the need might arise'. When this happens the specification rapidly becomes out of balance and inappropriately large. The inexperience of the specifier is immediately apparent to the estimator, and the tender is increased accordingly to take account of the risk of possible increased costs arising as a result of this lack of experience.

Clauses from libraries of standard landscape clauses must always be copied out in full, either by copy typing from a list of clause numbers prepared by the specifier, or (now more commonly) by use of a computerized 'mark up' copy to be returned to the originator to print out. It is not considered sufficient or good practice to incorporate (by reference) the full library of clauses, or to give only headings and code references to identify the various clauses. This latter practice would require the users – specifier, estimator and contract manager alike – to have to refer constantly to the source document which may or may not have been revised before or after the project specification was prepared, and increase the possibility of dispute and subsequent litigation arising in such instances.

It is now also possible to prepare a standard office library of landscape clauses based on a nationally accepted set of standard clauses in the same order of arrangement and coding, with one's own selection of the standard clauses and own special clauses inserted and coded accordingly. This is only practicable if suitable arrangements are made for the regular updating of the clauses, and obtaining the necessary licence where appropriate. However, in practice, it rarely causes problems; most computer agencies are able to offer this facility for even the simplest commercially available in-house microcomputers, such as those made by Apple, Commodore and IBM.

3.7 Watering

The specification for the provision of adequate water is essential for the establishment and rapid growth of plants. A requirement to provide 'sufficient water during the Defects Liability Period' is as inadequate as the similar provisions for temporary protection. Only a clause such as the following (also Figure 3.1) will ensure proper provision for watering in the contract documents:

The contractor is to water all trees, shrubs, grass and other planted areas at the time of planting in order that the entire tree pit or planted area is moistened to field capacity.

6401
The number of waterings required will depend on the weather conditions and the number to be allowed for will depend on the anticipated time(s) of planting. Insert number required, which may vary for different types of plant, or the words 'as measured'.

```
6401  WATERING: ensure that sufficient water is applied to
      maintain healthy growth of trees/shrubs/plants.
      Suggest to SO when watering may be required and when
      instructed carry out using a fine rose or sprinkler
      until full depth of topsoil is saturated.  Number of
      waterings to be allowed for:

      - - - - -
      - - - - -

6402  WATERING: spray crown of trees when in leaf during warm
      weather.  Carry out in the evening.
```

Fig. 3.1 An example of specification clauses to cover watering requirements. (Reproduced with permission of NBS Services Ltd.)

Watering shall be undertaken on subsequent occasions when directed by the landscape architect to ensure the establishment and growth of the plants.

To comply with this clause, the contractor is to allow for the provision of water, water carts or hoses with a fine rose attachment or sprinklers at normal mains pressure, allowed to run until the full depth of topsoil specified has reached a point where full absorption is achieved over the areas of all trees, shrubs and grass on each occasion. The contractor is to include and state in his tender the cost of compliance with this clause so that the cost of visits can be deducted in whole or in part if not required to be used.

It is important to then refer to this and the following item in the bills, schedules or tender summary, to ensure that it is included and priced for.

Should emergency legislation restricting the use of water during drought conditions be imposed, the contractor will be required to ascertain, before planting, the availability of and arrange the collection and application of second-class water by bowser or other means from an approved source, deliver to site and apply as specified (Figure 3.2).

During 'dry spells' water authorities are empowered to ban the use of water supplied by hosepipes for washing private cars and watering private gardens. It is important to note that this only applies to watering after practical completion

1151
Use this clause only when the planting is to be done by a nominated sub-contractor.

```
1151  WATER will be provided by the Main Contractor up to
      Practical Completion of the Main Contract Works and
      thereafter by the Employer, always subject to availa-
      bility of supply.

1201  DROUGHT CONDITIONS: if water supply is or is likely to
      be restricted by emergency legislation, inform SO
      without delay and ascertain availability and additional
      cost of second class water from a sewage works or other
      approved source.
      1.  If planting has not been carried out, do not do so
          until instructed.
      2.  If planting has been carried out, obtain instructions
          on supply of water.
```

Fig. 3.2 An example of specification clauses to cover action to be taken in drought conditions. (Reproduced with permission of NBS Services Ltd.)

and does not apply to landscape contracts to which the public has access. For private gardens after practical completion landscape contractors must therefore arrange for such watering as has been specified to be provided other than from a hosepipe.

It is in any case always prudent for landscape contractors to be provided with their own supply to be metered separately.

4

Landscape quantities

4.1 Measurement of landscape work on with- and without-quantities projects

The primary objective of any set of tender and subsequent contract documents is to ensure that everything required for the proper completion of the work is included by the contractor, and that tenders are submitted on a comparable basis (which is not to say that there may not be a wide range of different prices submitted by the individual tenders for the various elements of the work while the tender totals may not be dissimilar).

The secondary objective is to provide comparable rates for the landscape architect or quantity surveyor to value any variations to the contract and as a basis for valuations prepared in support of Interim and Final Certificates.

To achieve these two objectives it is essential on any project, large or small, to ensure that the tender and contract documents contain sufficient information for it to be priced by the contractor for this to be achieved. All standard forms of contract, therefore, contain provisions for the contractor to 'price the specification or schedules or schedule of rates' (JCLI, 2nd Recital).

Even when quantities are not provided by the employer it is advisable for specifications and tender summaries to state the overall areas of topsoil, grass and shrub borders, so that in the evaluation of tenders submitted it is certain that the areas included are comparable and that one tenderer has not included 60 square metres grass and 40 square metres shrub border, whereas another has taken 40 square metres grass and 60 square metres shrub border (both measured from the same drawing). To enter into a contract with such an error or discrepancy, albeit with the lowest tenderer, makes subsequent accurate financial control almost impossible.

Errors of this sort are most likely to arise when details at different scales are included on the same drawing. It is essential that the scales are clearly marked on each detail.

4.2 Schedules

Even the smallest project will need a schedule in one form or another. BS 1192: Part 1 defines a schedule as 'Tabulated information on a range of similar items' and gives separate tables for tree work, planting shrubs and climbers. (Tables 4.1–4.3 show examples of such schedules.)

Table 4.1 Example of tree work schedule (after BS 1192)

Tree No. and Tree Preservation Order (PTO) No.	Name	Protection	Work required	Cost*
1 (113717)	*Quercus robur*	Barrier type . . . erected all round. No nearer the bole than the canopy drip line.	Crown reduction to the South. Remove dead wood.	
2	*Fraxinus ornus*	Barrier type . . . erected where indicated on drawing no. . . .	Clean wound to the East; trim to live tissue and treat with approved wound sealant. Hand dig service duct trench. Roots greater than 20 mm not to be severed.	

*This column is optional.

Table 4.2 Example of tree planting schedule

Quantity	Name	Designation	Height/girth (where applicable*) cost	Root system	Planting location			Unit rate[†] £	Total cost[†] £
					A1	A2	A3		
3	*Robinia pseudoacacis*	Tall standard		Bare root	3				
5	*Betula pendula* 'Tristis'	Feathered	3.50 m (100 mm)	Bare root		2	3		

*Dimensions of standard trees as in BS 4043 and BS 3936: Part 1.
[†]These columns are optional.

Table 4.3 Example of planting schedule for shrubs and climbers

Quantity	Name	Root system/container volume	Height/ spread*	Planting location			Unit rate £	Total cost[†] £
				P1	P2	P3		
9	*Rhus typhina*	Container grown 2.5 L	900	–	3	6		
100	*Vinca minor*	Bare root	300 sp	–	30	70		

*When specifying spread use suffix sp.
[†]This column is optional.

4.3 Bills of Quantities

Traditionally **Bills of Quantities** are provided by the employer for inclusion in the contract documents on only the larger landscape contracts, many contracts being let on a 'drawings and spec' basis.

To provide a uniform basis for measuring building works a committee set up by the quantity surveyors and contractors in 1922 has, over the succeeding years, produced a series of standard methods of measurement, the latest being the seventh edition (SMM7) (Figure 4.1). This was a tremendous improvement on the previous edition with the woefully inadequate treatment of landscape works dismissed summarily in:

General rules

1. Introduction

1.1
This Standard Method of Measurement provides a uniform basis for measuring building works and embodies the essentials of good practice. Bills of quantities shall fully describe and accurately represent the quantity and quality of the works to be carried out. More detailed information than is required by these rules shall be given where necessary in order to define the precise nature and extent of the required work.

1.2
The rules apply to measurement of proposed work and executed work.

2. Use of the tabulated rules

Generally

2.1
The rules in this document are set out in tables. Each section of the rules comprises information (to be) provided, classification tables and supplementary rules. The tabulated rules are written in the present tense.

2.2
Horizontal lines divide the classification table and supplementary rules into zones to which different rules apply.

Classification tables

2.3
Within the classification table where a broken line is shown, the rules given above and below the broken line may be used as alternatives.

2.4
In referring to columns in classification tables the measurement unit column has been disregarded.

Fig. 4.1 SMM7 – Standard Method of Measurement of Building Works. (Reproduced with permission of the Standing Joint Committee for the Standard Method of Measurement of Building Works.)

2.5
The left hand column of the classification table lists descriptive features commonly encountered in building works. The next column lists further sub-groups into which each main group of items shall be divided and similarly the third column provides for further division. The lists in these columns are not intended to be exhaustive.

2.6
Each item description shall identify the work with respect to one descriptive feature drawn from each of the first three columns in the classification table and as many of the descriptive features in the fourth column as are applicable to the item. The general principle does not apply to Preliminaries in that it will be necessary to select as many descriptive features as appropriate from each column.

2.7
Where the abbreviation (nr) is given in the classification table the quantity shall be stated in the item description.

Supplementary rules

2.8
Within the supplementary rules everything above the horizontal line, which is immediately below the classification table heading, is applicable throughout that table.

2.9
Measurement rules set out when work shall be measured and the method by which quantities shall be computed.

2.10
Definition rules define the extent and limits of the work represented by a word or expression used in the rules and in a bill of quantities prepared in accordance with the rules.

2.11
Coverage rules draw attention to particular incidental work which shall be deemed to be included in the appropriate items in a bill of quantities to the extent that such work is included in the tender documents. Where the coverage rules include materials they shall be mentioned in the item descriptions.

2.12
The column headed Supplementary Information contains rules governing the information which shall be given in addition to the information given as a result of the application of rule 2.6.

2.13
A separate item shall be given for any work which differs from other work with respect to any matter listed as supplementary information.

Fig. 4.1 *(Contd)*

3. Quantities

3.1
Work shall be measured net as fixed in position except where otherwise stated in a measurement rule applicable to the work.

3.2
Dimensions used in calculating quantities shall be taken to the nearest 10mm (i.e. 5mm and over shall be regarded as 10mm and less than 5mm shall be disregarded).

3.3
Quantities measured in tonnes shall be given to two places of decimals. Other quantities shall be given to the nearest whole unit except that any quantity less than one unit shall be given as one unit.

3.4
Unless otherwise stated, where minimum deductions for voids are dealt with in this document they shall refer only to openings or wants which are within the boundaries of measured areas. Openings or wants which are at the boundaries of measured areas shall always be the subject of deduction irrespective of size.

3.5
The requirement to measure separate items for widths not exceeding a stated limit shall not apply where these widths are caused by voids.

4. Descriptions

4.1
Dimensions shall be stated in descriptions generally in the sequence length, width, height. Where ambiguity could arise, the dimensions shall be identified.

4.2
Information required by the application of rules 2.6 and 2.12 may be given in documents (e.g. drawings or specification) separate from the bills of quantities if a precise and unique cross reference is given in its place in the description of the item concerned. This rule does not allow the aggregation of a number of measured items which are otherwise required to be measured separately by these rules, except as provided by rule 9.1.

4.3
Headings to groups of items in a bill of quantities shall be read as part of the descriptions of the items to which the headings apply.

4.4
The use of a hyphen between two dimensions in this document or in a bill of quantities shall mean a range of dimensions exceeding the first dimension stated but not exceeding the second.

4.5
Each work section of a bill of quantities shall begin with a description stating the nature and location of the work unless evident from the drawn or other information required to be provided by these rules.

Fig. 4.1 (*Contd*)

4.6
Unless otherwise specifically stated in a bill of quantities or herein, the following shall be deemed to be included with all items:
(a) Labour and all costs in connection therewith.
(b) Materials, goods and all costs in connnection therewith.
(c) Assembling, fitting and fixing materials and goods in position.
(d) Plant and all costs in connection therewith.
(e) Waste of materials.
(f) Square cutting.
(g) Establishment charges, overhead charges and profit.

4.7
A dimensioned description for an item in the bill of quantities shall define the item and state all the dimensions necessary to identify the shape and size of the item or its components.

5. Drawn information

5.1
Location drawings:
(a) Block Plan: shall identify the site and locate the outlines of the building works in relation to a town plan or other context.
(b) Site Plan: shall locate the position of the building works in relation to setting out points, means of access and general layout of the site.
(c) Plans, Sections and Elevations: shall show the position occupied by the various spaces in a building and the general construction and location of the principal elements.

5.2
Component drawings: shall show the information necessary for manufacture and assembly of a component.

5.3
Dimensioned diagrams: shall show the shape and dimensions of the work covered by an item and may be used in a bill of quantities in place of a dimensioned description, but not in place of an item otherwise required to be measured.

5.4
Schedules which provide the required information shall be deemed to be drawings as required under these rules.

6. Catalogued or standard components

6.1
A precise and unique cross-reference to a catalogue or to a standard specification may be given in an item description instead of the description required by rules 2.6 and 2.12 or instead of a component drawing.

Fig. 4.1 (*Contd*)

7. Work of special types

7.1
Work of each of the following special types shall be separately identified:

(a) Work on or in existing building – see general rule 13.
(b) Work to be carried out and subsequently removed (other than temporary works).
(c) Work outside the curtilage of the site.
(d) Work carried out in or under water shall be so described stating whether canal, river or sea water and (where applicable) the mean Spring levels of high and low water.
(e) Work carried out in compressed air shall be so described stating the pressure and the method of entry and exit.

8. Fixing, base and background

8.1
Method of fixing shall only be measured and described where required by the rules in each Work Section. Where fixing through vulnerable materials is required to be identified, such materials are deemed to include those listed in rule 8.3 (e).

8.2
Where the nature of the base is required to be identified each type of base shall be identified separately.

8.3
Where the nature of the background is required to be identified the item description shall state one of the following:

(a) Timber, which shall be deemed to include manufactured building boards.
(b) Masonry, which shall be deemed to include concrete, brick, block and stone.
(c) Metal.
(d) Metal faced materials.
(e) Vulnerable materials, which shall be deemed to include glass, marble, mosaic, tiled finishes and the like.

9. Composite items

9.1
Notwithstanding the requirement of clause 4.2, work to be manufactured off site may be combined into one item even though the rules require items to be measured separately, provided the items in question are all incorporated into the composite item off site. The item description shall identify the resulting composite item and the item shall be deemed to include breaking down for transport and installation and subsequent re-assembly.

Fig. 4.1 (*Contd*)

10. Procedure where the drawn and specification information required by these rules is not available

10.1
Where work can be described and given in items in accordance with these rules but the quantity of work required cannot be accurately determined, an estimate of the quantity shall be given and identified as an approximate quantity.

10.2
Where work cannot be described and given in items in accordance with these rules it shall be given as a Provisional Sum and identified as for either defined or undefined work as appropriate.

10.3
A Provisional Sum for defined work is a sum provided for work which is not completely designed but for which the following information shall be provided:

(a) The nature and construction of the work.

(b) A statement of how and where the work is fixed to the building and what other work is to be fixed thereto.

(c) A quantity or quantities which indicate the scope and extent of the work.

(d) Any specific limitations and the like identified in Section A35.

10.4
Where Provisional Sums are given for defined work the Contractor will be deemed to have made due allowance in programming, planning and pricing Preliminaries. Any such allowance will only be subject to adjustment in those circumstances where a variation in respect of other work measured in detail in accordance with the rules would give rise to adjustment.

10.5
A Provisional Sum for undefined work is a sum provided for work where the information required in accordance with rule 10.3 cannot be given.

10.6
Where Provisional Sums are given for undefined work the Contractor will be deemed not to have made any allowance in programming, planning and pricing Preliminaries.

11. Work not covered

11.1
Rules of measurement adopted for work not covered by these rules shall be stated in a bill of quantities. Such rules shall, as far as possible, conform with those given in this document for similar work.

Fig. 4.1 (*Contd*)

12. Symbols and abbreviations

12.1

The following symbols and abbreviations are used in this method of measurement:

m	=	metre
m^2	=	square metre
m^3	=	cubic metre
mm	=	millimetre
nr	=	number
kg	=	kilogramme
t	=	tonne
h	=	hour
p c sum	=	Prime Cost Sum
prov sum	=	Provisional Sum
$>$	=	exceeding
$\geq$	=	equal to or exceeding
$\leq$	=	not exceeding
$<$	=	less than
%	=	percentage
–	=	hyphen (see rule 4.4)

12.2

Cross references within the classification tables are given in the form:

Work Section number	:	Number from first column	.	Number from second column	.	Number from third column	.	Number from fourth column

Example:

D20: 2. 2. 2. 1

Excavation and filling

Excavating

To reduce levels

Maximum depth $\leq$ 1.00m

Commencing level stated where $>$ 0.25m below existing ground level.

12.3

An asterisk within a cross reference represents all entries in the column in which it appears.

12.4

The digit 0 within a cross reference represents no entries in the column in which it appears.

13. Work to existing buildings

13.1

Work to existing buildings shall be so described. Such work is defined as work on, or in, or immediately under work existing before the current project.

Fig. 4.1 (*Contd*)

13.2
The additional rules for work to existing buildings are to be read in conjunction with the preceding rules in the appropriate Work Sections.

13.3
A description of the additional Preliminaries/General conditions which are pertinent to the work to the existing building shall be given, drawing attention to any specific requirements due to the nature of the work.

14. General definitions

14.1
Where the rules require work to be described as curved with the radii stated details shall be given of the curved work including if concave or convex, if conical or spherical, if to more than one radius and shall state the radius or radii.

14.2
The radius stated shall be the mean radius measured to the centre line of the material unless otherwise stated.

Fig. 4.1 (*Contd*)

Clause D44 – Landscaping

Soiling, cultivating, seeding, treating top surface, fertilizing and turfing to surfaces shall be given separately in square metres, stating the thickness of soil and the quantity per square metre of the seed or fertilizer. Any requirements for watering, weeding, cutting, settlement, re-seeding and the like shall be stated.

Against the often-repeated complaint – that 80% of a surveyor's and estimator's time is spent measuring and pricing 20% of the value of a contract – it had to be admitted that the particular needs of landscape contracting required more than this if contractors were to be given sufficient information in a standard format and sequence for them to submit an accurately and comparably priced tender. Irrespective of the quality of the rules themselves a standard method is essential to establish the minimum information necessary to define the nature and extent of the work, items not listed being deemed not to have to be included in the tender, and any departures from the standard measurement conventions having to be specifically drawn to the attention of the estimator.

To overcome the previous deficiency the SMM Committee in 1987 prepared a new edition (SMM7) (Figure 4.1) rearranged in the 'Common Arrangement' format giving far more information to the contractor's estimator at tender stage and enabling the client's quantity surveyor to price his valuations for interim certificates and variations much more accurately. The requirements of the landscape industry were submitted to the committee at the drafting stage of the new edition and have now been incorporated making the earlier 1977 BALI supplementary rules for their *Method of Measurement for Soft Landscape Works* published by JCLI in 1978 obsolete.

Q30 Seeding/Turfing

INFORMATION PROVIDED					MEASUREMENT RULES	DEFINITION RULES	COVERAGE RULES	SUPPLEMENTARY INFORMATION
P1 The following information is shown either on location drawings under A Preliminaries/General conditions or on further drawings which accompany the bills of quantities: (a) the scope and location of the work								
CLASSIFICATION TABLE								
1 Cultivating	1 Depth stated		m²	1 Weeding, details stated 2 Cutting, details stated 3 Preparatory work, details stated		D1 Types of surface applications include herbicides, selective weedkillers, peat, manure, compost, mulch, fertilizer, soil ameliorants, sand and the like	C1 Cultivating is deemed to include the removal of stones C2 Surface applications are deemed to include working in if required C3 Seeding is deemed to include raking or harrowing in and rolling C4 Cutting is deemed to include edge trimming	S1 Timing of operations S2 Method of cultivating and degree of tilth S3 Kind, quality, composition and mix of materials S4 Method of application S5 Method of securing turves
2 Surface applications	1 Type and rate stated							
3 Seeding	1 Rate stated							
4 Turfing								
5 Turfing edges of seeded areas	1 Width stated							
6 Protection	1 Temporary fencing			1 Duration and ultimate ownership, details stated	M1 Protective temporary fencing is only measured here where specifically required and then in accordance with Section Q40			

Q31 Planting

INFORMATION PROVIDED					MEASUREMENT RULES	DEFINITION RULES	COVERAGE RULES	SUPPLEMENTARY INFORMATION
P1 The following information is shown either on location drawings under A Preliminaries/General conditions or on further drawings which accompany the bills of quantities: (a) the scope and location of the work								

CLASSIFICATION TABLE								
1 Cultivating	1 Depth stated		m²	1 Weeding, details stated 2 Fallowing, details stated		D1 Types of surface applications include herbicides, selective weedkillers, peat, manure, compost, mulch, fertilizer, soil ameliorants, sand and the like D2 BS size designations include standard, advanced nursery stock or semi-mature trees D3 Young nursery stock includes seedlings, transplants and whips D4 Removing surplus excavated material means removing from site unless otherwise described	C1 Cultivating is deemed to include the removal of stones C2 Surface applications are deemed to include working in if required C3 Items include for excavating or forming pits, holes or trenches, refilling, watering in, removing surplus excavated material and labelling C4 Refilling is deemed to include all necessary multiple handling C5 Planting in cultivated or grassed areas prepared by others is deemed to include all necessary reinstatement	S1 Timing of operations S2 Method of cultivating and degree of tilth S3 Kind, quality and composition of materials S4 Size and type of pits, holes and trenches excavated or formed S5 Type of supports and ties S6 Special materials for refilling S7 Labelling
2 Surface applications	1 Type and rate stated							
3 Trees	1 Botanical name	1 BS size designation and root system stated	nr	1 Planting in cultivated or grassed areas prepared by others, details stated 2 Initial cut back, details stated 3 Supports and ties 4 Refilling with special materials, details stated 5 Watering, details stated				
		2 Girth, height and clear stem and root system stated						
4 Young nursery stock trees 5 Shrubs		1 Height and root system stated	nr					
6 Hedge plants		1 Height stated	nr					
		2 Height, spacing, number of rows, and layout stated	m					
7 Herbaceous plants		1 Size stated	nr					
		2 Size and number per m² stated	m²					
8 Bulbs, corms and tubers		1 Size stated	nr					
			kg					
9 Mulching after planting	1 Around individual plants	1 Thickness and area stated	nr	1 Tree spats, details stated				S8 Type of mulch, time and method of application
	2 Beds	2 Thickness stated	m²					
10 Protection	1 Tree guards	1 Dimensioned description	nr					S9 Type of tree guard and method of fixing S10 Type of spray and rate of application S11 Type of wrapping and chemical application
	2 Anti-desiccant sprays	2 Height and girth of tree or spread of plant stated						
	3 Wrapping	3 Height of wrapping and girth of tree stated						
	4 Temporary fencing			1 Duration and ultimate ownership, details stated	M1 Temporary fencing is only measured here where specifically required and then in accordance with Section Q40			

Fig. 4.2 SMM7/Q30 Seeding/Turfing and Q31 Planting. (Reproduced with permission.)

The landscape industry now has a set of rules in the same Common Arrangement format as the building industry so that Bills of Quantity can be cross-referenced to drawings and specifications. An example giving details of Seeding/Turfing and Planting is found in Figure 4.2.

In addition to the measured items it is often also useful to include within the Bills of Quantities Schedules prepared in accordance with BS 1192 listing, for example, work to existing trees, numbers, species and size of new trees, shrubs and climbers, examples of which are given in Tables 4.1, 4.2 and 4.3.

5

Landscape lump sum contract conditions

This chapter attempts only to summarize the salient features of the lump sum contracts for landscape works. (For a full commentary see *The Shorter Forms of Building Contract*, Granada Publishing, London, 1984.)

5.1 The need for a Standard Form

Landscape contracts, like building contracts, differ from other contracts in that they are all 'work and material' contracts. They are not contracts solely for the sale of goods at common law, nor are they subject to the Sales of Goods Act 1979 even though the title in such materials as are required ultimately passes to the building owner. Equally, they are not contracts for services alone and are not therefore covered by the Supply of Goods and Services Act 1982, Part II.

Landscape contracts do not have to be in any special form, they can be – and often are (for the smaller project) – by word of mouth. This inevitably causes problems, with misunderstandings as to what actually was agreed or what was to be done when circumstances arise that the parties have neglected to cover.

Frequently, 'contracts' are entered into by exchange of letters – the offer and its acceptance. While the terms of a standard contract can be incorporated by reference, e.g. *Modern Buildings v. Limmer & Trinidad (1975)*, this can also cause problems when the contract contains alternative provisions, e.g. fixed price or fluctuations. Problems also arise as to which edition was current at the time the contract was entered into when a contract has been subsequently amended.

If a new contract is drafted for each new project, or a local authority drafts its own, contractors will be suspicious of it and price accordingly. Also, by the Unfair Contract Terms Act 1977, some terms of a contract drafted unilaterally may be construed by the courts to be unenforceable and, where ambiguities arise, will be held to be *contra proferentum*, i.e. the meaning most favourable to the contractor will be adopted by the court. The courts have held, however, that where a contract has been agreed by all sides of the industry, this will not apply and it will not be considered an 'employer's' contract nor come within the provisions requiring endorsement by the Office of Fair Trading.

Fortunately for landscape contracts a Standard Form exists in the JCLI Standard Form of Agreement for Landscape Works (see Appendix C), with contract conditions covering the majority of situations considered likely to arise. Certain employers may, of course, find it necessary to annex their own special standard requirements but these should always be kept to the minimum.

5.2 The JCLI Standard Form of Agreement for Landscape Works

The Landscape Institute had published a Standard Form of Agreement for Landscape Work based on the JCT63 Form of Building Contract without Quantities edition in 1969, followed by the 'with Quantities' version in 1973. The two forms were not however kept up-to-date with any of the subsequent JCT amendments. In April 1978 the JCLI, with the co-operation of the LI, prepared a new Form; this, with the additional clauses necessary for landscape work, was almost identical in content and format to the then current 1968 JCT Form of Agreement for Minor Building Works. This had been agreed with the JCT in order to reduce the proliferation of standard forms and to ensure the widespread acceptance of the JCLI Form by local authority and county councils who were also represented on the JCT. Since the designation of 'landscape architect' is not protected as is the use of the name 'architect' by the Architects Registration Acts, there is no need for the alternative 'supervisory officer' to be added to the word 'landscape architect' and the Form as printed is therefore equally suitable for unamended use by landscape scientists and managers.

The Form was revised in April 1981 to bring it into line with the new clauses and sequence of JCT 1980 Minor Works edition and was reprinted with minor corrections in April 1982 and again in 1985 to include the option of Bills of Quantities and to enable the form to be used in Scotland. It is now the recommended form of contract for all public local authority and private sector landscape contracts (landscape works executed as subcontracts are dealt with in the following chapter).

The 1981 edition provides the requirement for the contractor to price any Schedules or schedules of rates included as contract documents in the 1st and 2nd Recitals, so that the Form is suitable for both those contracts with quantities and those with only drawings and specifications. It also provides for the option of naming any quantity surveyor the employer intends to appoint, e.g. for the valuation of variations, but, as in the JCT Minor Works Contract, does not specify any particular duties and also specifically excludes from the contract sum the value added tax that is always due from the employer on landscape contracts.

5.3 Arbitration

It is essential to realise that since *Sutcliffe v. Thakrah* the landscape architect is no longer considered a 'quasi- arbitrator' since he is an employee of the client with only the duty under the contract to act 'fairly and reasonably'. Only when the contractor gives notice that he considers this to be no longer the case is a 'dispute' considered to have arisen. Then the provisions of the arbitration clause for the appointment of an independent arbitrator under the 1950 and 1979 Arbitration Acts come into force. The effect of this is therefore that the landscape architects' decisions are no longer inviolate and can always be challenged if the landscape contractor feels sufficiently aggrieved and if he considers the amount or principle involved to justify the not-inconsiderable costs of arbitration proceedings.

Regrettably, from time to time the parties to the contract – namely the employer, or the landscape architect on his behalf, and the contractor – cannot reach agreement on some dispute that has arisen between them, such as the valuation of a particular variation or an extension to the contract period for completion of the works. Most contracts provide for this eventuality by the inclusion of a clause providing for disputes to be submitted to one or more impartial persons to settle the matter.

The JCLI Contract is no exception and sets out in Article 4 for such disputes to be submitted to the binding decision of an arbitrator agreed between them or should this not be possible, appointed by the President or Vice President of the Landscape Institute. Under these circumstances the courts are reluctant to interfere and hear the case, unless there is some specific question of law involved. Such arbitrations are governed by the Arbitration Acts of 1950 and 1979, when the arbitrator's decision, published in the form of an Award, is final and binding on both parties and without the right of appeal. Arbitrations usually take place

Dear Sirs,

Whereas disputes and differences have arisen between you and ourselves within the meaning of Article 4 of the Contract dated .... between .... and .... I here give you notice of our intention to have such differences settled by arbitration in accordance with the provisions of the Article.

Yours faithfully,

Fig. 5.1 Example letter setting arbitration procedures in motion.

after practical completion. Should this situation arise, a letter on the lines shown in Figure 5.1, setting the necessary procedures in motion, should suffice.

An arbitration will usually take the following form:

1. After a preliminary meeting between the parties the Arbitrator will issue directions for the exchange of 'pleadings' and the arrangements for the place, date and time of the hearing.
2. When the parties have submitted to each other in writing a short summary of their claim, the defence and counter claim, reply, any further or better particulars, and have inspected the supporting documents and evidence, the Arbitrator fixes a date and time for the hearing.
3. At the hearing, each side presents his case with full supporting evidence, witnesses and experts. A hearing will normally last two or three days although hearings of six months or even a year are not unknown.
4. After the close of the hearing the Arbitrator draws up his Award, giving his decision in writing, and directs which of the parties is to pay their own and the other's costs.

Some of the advantages for landscape contracts of submitting disputes to arbitration rather than submitting them to the courts, particularly if the dispute is on technical matters, are:

1. The hearing is private with only the parties to the dispute, their lawyers, witnesses and experts present.
2. The time and place of the hearing can be arranged to suit the convenience of all concerned.
3. If the parties so wish, a decision can be arrived at more quickly and cheaply than through the courts.
4. The person chosen as Arbitrator will usually be fully conversant with the technical aspect of the problem.

The case of *Oram Builders Ltd v. M.J. Pemberton* in 1985 established the right of an Arbitrator appointed on similar terms to those in the JCLI Form to open up and review an architect's certificate whereas a High Court Judge could not except in the unlikely event of his having been appointed to settle the matter as an arbitrator and not as Official Referee or as a Judge.

A similar case was *Crestar Ltd v. Carr and Ano* in the same year when a contract was completed four months late and after a further three months an architect issued a further certificate for a further amount increasing the cost of the work by a further 50%. It was also held that the arbitration clause could be held to apply notwithstanding the employer's obligation to honour certificates within 14 days.

5.4 The Conditions

Having left space blank at the top of the contents page of the contract, for the parties to enter into the contract under seal (should they desire to extend the statutory period of Limitation from 6 to 12 years) the actual Conditions of the contract are contained on four A4 pages under eight headings. (Appendix C contains a specimen copy of the JCLI Standard Form.)

1 The intentions of the parties

1.1 The **contractor** is obliged to carry out and complete the works in a good and workmanlike manner, only being relieved of his obligation to comply with the contract documents where the landscape architect has reserved to himself the approval of the quality of certain materials and workmanship. (This is of course irrespective of the landscape architect's responsibilities to his client to monitor the work of the contractor.)

1.2 The **landscape architect** (who is not a party to the contract) is made specifically responsible for supplying, on the employer's behalf, any necessary further information, issuing certificates and confirming all instructions in writing.

1.3 With the advent, in the Standard Method of Measurement for Building Works, of fully detailed Rules for the measurement of Landscape Works it was possible to provide for the use of the Form on Landscape Contracts where Bills of Quantities were provided by the employer to the contractor for tendering purposes and then included in the contract documents for the valuation of variations and interim certificates.

2 Commencement and completion

2.1 Space for inserting the date given in the tender documents on which the work **may** be commenced (there is no obligation on the contractor to do so) is left, as is space for the date by which the contractor is obliged to complete the work.

2.2 The landscape architect is responsible for **extending** in writing the period for completion of the works by a reasonable amount of time for reasons which the contractor has notified him as having delayed completion (not progress) and which were not under his (the contractor's) control. Most landscape architects will consider the 'relevant events' in JCT80 clause 25 as being reasons beyond the control of the contractor. There is however no requirement for the landscape architect to justify what he considers to be reasonable nor to quantify and total each individual reason except to calculate any Direct Loss and Expense due to disturbance in the progress of the works by the Employer and to which the contractor is entitled under Clause 3.6. This clause now specifically includes for extensions for delays in completion caused by compliance with the landscape architect's instructions and excludes delays within the control of the contractor.

2.3 Pre-estimating the amount of **damages for non-completion** (never, it must be remembered, to be punitive) in this way saves the necessity of proving in court the amount of damages suffered – which would otherwise be required if the damages were left at large. In assessing the amount of such damages, allowances must be made for:

'notional loss of interest on capital'
'inconvenience or any actual consequential loss'
'additional professional fees'.

It is geneally accepted that the amount of notional loss of interest per week is arrived at by such a formula as:

$$\frac{\text{Contract sum}}{\text{52 weeks}} \times \frac{\text{bank rate}}{100} = £\ldots \text{ per week}$$

To this should be added any consequential loss (rare on landscape projects) together with a pre-estimate of the weekly cost of any additional professional fees. The clause now specifically provides also for the deduction of damages by the Employer from interim certificates since it was possible for the particularly astute contractor to insist that he be paid the full amount of the certificate before he could refund the damages to which the Employer was entitled.

2.4 The **completion date** to be certified by the landscape architect is that when in his opinion the works have reached 'practical completion'. Not 'practically' (i.e. almost) complete but 'practical completion of the works'. This is generally accepted as being when the works are sufficiently complete for them to be safely and conveniently used for the purpose for which they have been designed. No strict definition of these words, in the restricted use to which they are put, has ever been laid down in the courts (and it is hoped never will be since were this to happen it would restrict the judgment of the landscape architect), every particular situation being considered on its merits.

2.5 **Defects** in all items of work – other than the failure of trees, shrubs, grass and other plants – are to be made good at the contractor's expense within three months (or such other period as is stated in the tender documents and inserted here) and the landscape architect is to certify when in his opinion these have been made good. This clause now provides also for defects in work executed as a Variation in addition to that included in the original contract document.

2.6 **Partial possession** is not provided for in the JCT Form since it was not considered likely to arise in the type of project for which the form was intended. On landscape projects however, it was appreciated that circumstances could arise such as when the work was completed in all respects, except that the seeding of grass areas would have to be postponed until such time as its germination could be ensured, or the planting of tree clumps postponed until bare root stock could be safely lifted. Under these circumstances, rather than issue a certificate of practical completion and exclude the grass or trees as 'an outstanding item' under this clause, it is possible in landscape contracts to leave part of the site in the possession of the contractor. He will therefore remain responsible for its security and protection from malicious damage etc., and the employer (with the consent of the contractor), for reasons not envisaged at the commencement of the contract, may take possession and have the use of and pay for that part of the site that has been finished, release half the retention on the part handed over and start the Defects Liability Period from the date that the part was taken over.

2.7 **Plant failures** prior to practical completion of the works are always to be replaced by the landscape contractor entirely at his own cost. Those found to be missing as a result of theft or malicious damage are dealt with under the separate clause, 6.5A or 6.5B.

JCLI Practice Note 3 deals with plant failures at practical completion as follows:

> All plants which have failed prior to Practical Completion or within six weeks of the first leafing out in the case of trees, shrubs or grass at the first cut and roll whichever is the later, are to be replaced by the Contractor entirely at his own expense unless it arises from theft or malicious damage after Practical Completion.

At an appropriate time, the contractor and landscape architect should meet to establish the extent of any defects and failures. The landscape architect will then prepare a schedule of these defects and send it to the contractor within fourteen days. The contractor will inform the landscape architect when the work has been rectified.

After practical completion, plant failures are dealt with in one of two ways.

(a) If the upkeep of the site during the Defects Liability Period, inserted in the tender and contract documents for the various types of plants, is to be undertaken by the landscape contractor then he is made responsible for the replacement (at his own cost) of any of them that have died or are found to be dying while he has been responsible for their maintenance. The landscape architect is required to certify at the end of the period when this replacement has been satisfactorily completed.

(b) Alternatively, if the employer is responsible for the upkeep after practical completion, then the contractor is relieved of further responsibility should any plants subsequently be found to be dead or dying, since injudicious or inadequate maintenance by the employer could clearly have resulted in the early demise of the plants concerned. The employer will have to make his own arrangements and pay for the replacements whether through the initial contractor or another.

In early drafts of the contract the above word 'upkeep' was used instead of maintenance, it being considered this would differentiate it from the maintenance of the hard landscape and buildings; but it was decided by the LI that more common usage used to describe the Defects Liability Period should prevail.

3 Control of the works

3.1 **Assignment** of the whole contract either by the employer or by the landscape contractor without the consent of the other is precluded by this clause (which did not appear in the 1978 edition but was added in 1981 to conform with MW80).

3.2 **Subcontracting** without the consent of the landscape architect (which shall not be unreasonably withheld) is similarly precluded. It should be noted that there is no provision in this Form for work to be carried out concurrently on the site by others directly employed by the employer.

3.3 The contractor is obliged to keep a **competent person** in charge on site at all reasonable times, who is specifically authorized to receive instructions on the contractor's behalf. There is no provision for the right of access to the works either for the architect or a clerk of works. This is an implied condition, since the landscape architect could not otherwise carry out the various duties required of him in the conditions in the contract.

3.4 The landscape architect may **object to the employment** of certain individuals on site (not their employment by the contractor, which might conflict with the Employment Protection Act of 1975), should he consider them to be incompetent, or their behaviour to be offensive.

3.5 The landscape architect is empowered to issue variations and **instructions**, which by implication include those concerning opening up and the removal of defective work. These must be in writing. In addition, the 1981 edition carries the option for the employer to instruct others to do any work which the contractor has neglected to do in compliance with an instruction, and to deduct the cost from any money due to the contractor.

3.6 **Variations** may be ordered by the landscape architect (and are therefore legally binding on the employer to pay for them), to add, omit or change the works or the order or period when they are to be carried out, and for these to be valued by the landscape architect (not the contractor) on a fair and reasonable basis using the prices referred to in the First and Second Recital.

There is also the proviso, which does not exist in the JCT Minor Works Contract, that if any omission substantially varies the scope of the work, this shall take into account the effect on the cost of the remainder of the work. There is also the option for the landscape architect and the contractor to 'agree' the price of a variation prior to carrying it out – this is always to be recommended, if possible. The clause now includes also for the reimbursement of the contractor of any direct loss and expense to which he has been put by reason of disturbance of the progress of the work due to instructions of the Employer or Landscape Architect. This was previously covered in a separate clause 3.9 now deleted.

3.7 **Provisional and PC sums** are defined in SMM.

Provisional sums are for work or costs which cannot be entirely foreseen, defined or detailed at the time of issuing the tender documents. Prime cost (p.c.) sums are those for work by a nominated subcontractor, statutory authority, public undertaking or materials or goods to be obtained from a nominated supplier, the sum to be exclusive of any profit required by the general contractor, for which provision should be made for it to be included in the tender.

It is recommended that a clause such as the following be used:

Include the p.c. (provisional) sum of £ . . . . . . for . . . . to be executed by a specialist subcontractor selected by the landscape architect. Add for profit and attendance.

Additionally, if the identity of the subcontractor proposed is known at the time of the invitation to tender, his name should be inserted and equally important when appropriate the period of time the subcontractor needs to carry out and complete the work. This eliminates the possibility of any future claim from the main contractor. When the identity of a subcontractor is not known to him at the time of tender he has no way of finding out how long he needs to allow for the execution of the subcontract work.

There is no provision in the form for any cash discount from a nominated subcontractor or supplier, nor for any direct payment by the employer in the event of default by the main contractor.

Should it be possible to draw and/or describe the work to be carried out or supplied by specialists sufficiently accurately for it to be included in the tender and contract documents for it to be priced, it is often preferable to include the names of those firms the contractor is restricted to use, by including a clause such as the following.

The landscape work, in accordance with this specification and drawings no. . . ., shall be carried out by one of the following.

1 . . . .
2 . . . .
3 . . . .
4 . . . .
5 . . . .

or other landscape subcontractor approved prior to the submission of tenders.

There are no Standard Forms of subcontract that the main contractor is obliged to use or that is available for use for nominated, named or 'domestic' (i.e. chosen entirely at the discretion of the main contractor under clause 3.2) subcontractors in connection with the JCLI Form; that included in Appendix B may be appropriate, landscape architects and contractors must however ensure that it meets their own requirements before adopting it.

3.8 **Objections to a nomination** may be made by the main contractor but this clause will not apply if the identity of the subcontractor(s) has been made known to him at the time of tender in accordance with the previous clause. This clause does not however exist in the JCT Minor Building Works Form.

4 Payment

4.1 The correction of any inconsistencies that may be discovered, in or between the contract documents, is acknowledged in the Contract and is provided for with the simple statement that they shall be corrected and, if this involves a variation, valued as such, leaving every case to be treated on its merits. The question of precedence is dealt with by the statement that the contract conditions shall prevail over all others, leaving by implication the accepted convention that drawings take precedence for location, Bills for quantity, and the specification for quality.

4.2 **Progress payments** are provided for: 'when requested by the contractor at intervals of not less than 4 weeks from the date of commencement' (the actual date, **not** the one in the contract after which the works **may** be commenced) without any limitation on the minimum amount claimed. The landscape architect has only to include for work 'properly' executed. This does not relieve the contractor of his obligations under clause 1.1 to complete the works in accordance with the contract documents nor place any contractual obligation on the landscape architect to ensure that he does so. It does however enable him to deduct the value of any work he considers not to be in accordance with the contract, from an Interim Certificate. As they are continuing valuations, any work included in a previous certificate and subsequently found to be defective can be also deducted from the Certificate.

There is no specific provision for the removal of defective work; when this arises the landscape architect issues the necessary instructions under clause 3.5 and, in the event of non-compliance, the employer may arrange for the work to be carried out by others.

The clause also provides for unfixed materials on site to be included in Interim Certificates notwithstanding the fact that legally possession does not otherwise pass to the employer until they are fixed. To prevent problems of retention of title (such as those which arose to the Romalpa and Dawber Williamson cases, in neither of which was there any contract between the supplier of the goods in question and the employer) it is always advisable to include in the contract documents a clause such as the following.

At the time of each valuation the contractor is to disclose to the landscape architect which of the unfixed materials and goods on site are free from, and which are subject to, any reservation of title inconsistent with the passing of property as required by clause 4.2 of the Conditions of Contract, together with their respective values. When requested provide evidence of freedom from reservation of title. © NBS Services Ltd.

This is particularly important in the case of plants delivered to site and temporarily stored pending planting in their final locations. Equally, problems may arise with materials stored 'off site' until required, e.g. in the landscape contractor's nursery. In these instances they must be both properly insured and clearly identified if the landscape architect is to be asked to use his discretion and include them in an Interim Certificate.

Retention is provided by the deduction of 5% from each certificate but without any requirement for it to be identified and held in trust for the contractor. Amounts previously certified (not paid) are also to be deducted from the valuation for each certificate and the certificate is to be honoured within 14 days of issue (not receipt). It is now usual for the certificate to be forwarded to the client direct with a copy sent concurrently to the contractor (see Chapter 9).

4.3 **Penultimate Certificates** are to be issued within 14 days of practical completion, releasing half the amount of the retention and based on the contract sum adjusted to include the value of any variations and the adjustment of p.c. and provisional sums. For this reason if the contractor is to be responsible for the subsequent upkeep of the landscape work this should be valued separately (Chapter 9) and not be included in the main contract sum.

For the same reason there is provision in this clause for the amount of retention to be increased if the amount of plant failures are considered to be in excess of 10%; the amount retained is to be adjusted accordingly by increasing the retention by the amount of the Bill rates for the defective items. This Certificate is also to be honoured within 14 days of issue.

4.4 The **Final Certificate**. Under both JCT and JCLI Forms the contractor is obliged to provide all the financial evidence the landscape architect needs to complete the final valuation, within three months (or such other period as is inserted) of the Certificate of Practical Completion. The landscape architect must then issue his Final Certificate within the following 28 days **provided** the Defects Liability Period has expired, and provided the landscape architect has certified that all defective items have been made good in accordance with clause 2.5.

It must be emphasized that in this instance, although the architect has a duty to act fairly and has a duty to his client not to over-certify, he must not be subjected to improper pressures from his client to withold the certificate (*Hickman v. Roberts*, 1913). Nor is he immune from charges of negligence in the issue of a final certificate, it having been held since 1974 (*Sutcliffe v. Thakrah*) that he is no longer a 'quasi-arbitrator' as had been thought to have been the case since *Chambers v. Goldthorpe* in 1901.

The duty of the landscape architect to issue certificates is clearly laid down in the conditions of the contract. If he does not do so at the proper time the landscape contractor's remedy is to submit the matter to an arbitrator appointed by the current President of the Landscape Institute. This need not be an expensive or lengthy procedure and if any payment awarded is still not forthcoming it can be enforced by the courts.

4.5 **Contributions, levies and tax changes**. If any unforeseen changes are suddenly imposed by the Government they could impose considerable burdens on a contractor since, under a fixed-price contract, he could in no way pass them on to the employer. This eventuality is therefore covered by Part A of the supplementary Memorandum and incorporated by reference in clause 4.5. Since this covers only 'work people', and therefore excludes site and head office supervisory staff, levies and taxes on materials are covered by an additional percentage to be added to cover the cost of these items: 20% is generally considered appropriate for contracts in excess of 12 months, and 10% for contracts of 12 months or less. There is the option of deleting the whole clause if the contract is of such a limited duration as to make it inappropriate.

4.6 **Fixed price/fluctuations** Many landscape projects are of a duration of 12 months or less, when it is reasonable for a contractor to include in his pricing any change in the cost of labour, materials, plant or other resources subsequent to the submission of his tender. There are many other instances, such as when planting has to be postponed until the beginning of the next planting season or when the needs of the project are such that it has to be spread over two seasons from the outset, and it is then prudent to allow the contractor to recover the cost of any increase that may arise. This is done by using category 48 of the Board of Trade (NEDO Formula) monthly index of costs of landscape work in accordance with Part D of the Supplementary Memorandum and Formula Rules. These are incorporated by reference in the alternative clause 4.6B having deleted clause 4.6A.

Where provided for in the tender documents, the **recovery of unforeseen increased costs** arising during the progress of landscape works is achieved by the use of the 'Formula Rules' (clause 4.6B of the JCLI Form of Contract).

The Government's decision on the use of the Formula Method of calculating fluctuation in public sector work (DoE Circular 158/73, Welsh Code 313/73 and Scottish Office 2/1974, the preparation of cost indices prepared for NEDO by the EDC for Building) resulted in the publication of the HMSO booklet *Price adjustment formulae; guide to application and procedure (series 2)* and the *Description of indices*, using indices compiled and maintained by the Property Services Agency of the DoE as monthly bulletins.

The Formula Rules for work category indices (series 2) published by the Joint Contracts Tribunal are obtainable from the RIBA, the RICS and the BEC. They set out the method of adjusting the contract sum at each interim valuation by adjusting the value of work included in that valuation in accordance with the movement of the index numbers in the monthly bulletins (series 2) referred to above, with a practice note (33) and worked examples giving guidance on the application of the rules for the Standard Form of Building Contract.

With regard to the JCLI Standard Form for Landscape Works this is covered in Category 48 (series 2) 'soft landscaping' and is included in the list of indices published monthly. These are available from HMSO and are also published monthly in the building business section of *Building* magazine.

The application of the index follows the normal lines, e.g.

Increase

$$= \frac{\text{Valuation month index} - \text{base month index} \times \text{value of work done in month}}{\text{Base month index}}$$

Costs are based on June 1976 = 100.

5 Statutory obligations

5.1 **Statutory obligations, notices, fees and charges** are the responsibility of the contractor, whether specifically referred to in the contract documents or not. If the contractor finds anything that he considers will involve him in a contravention, the contract requires him to notify the landscape architect. Should he comply with the contract documents and a contravention is subsequently discovered, the contract provides for the cost of compliance to be recovered from the employer.

5.2 **Value added tax** on landscape work is nearly always due at the standard rate. Customs and Excise Note 708 specifically includes 'site restoration' which is defined as 'the clearance of rubble, levelling the land, the application of topsoil and the laying of grass and simple paths, the planting of trees or shrubs or elaborate ornamental work apart from new building work which was otherwise zero rated'. The contract sum being exclusive of value added tax, this is collected by the contractor separately from the employer, in accordance with Part B of the Supplementary Memorandum. It also provides the necessary procedures for the employer (through the contractor) to challenge any tax which he considers to have been incorrectly imposed.

5.3 **Statutory tax deduction scheme** legislation makes the employer liable for any tax on the income of a self-employed person in the building industry which he may not have paid. It is therefore essential that employers, or landscape architects on their behalf, obtain copies of, or sight of, the current certificates of exemption from all the contractors and subcontractors they intend to employ, and to establish the necessary procedures at the outset of the contract so that they are not subsequently held responsible for the unpaid tax of the firms they employ.

5.4 This clause number is no longer used having previously been in the Fair Wages clause required for Local Authority Contracts before it was rescinded.

5.5 **Prevention of corruption**. By the use of the words 'shall be entitled to' this clause allows the employer the option of cancelling the contract and recovering the amount of any loss which he may have incurred as a result of the contractor having made any gift or consideration covered by the various Prevention of Corruption and Local Government Acts.

6 Injury, damage and insurance

6.1 **Injury to, or death of, persons** arising from carrying out the work is the sole responsibility of the contractor who has not only to indemnify the employer, but also to take out sufficient insurance to ensure that he has the financial resources to meet any claim which may arise (even after the work is finished) unless it arises from any act or neglect of the employer or any person for whom the employer is responsible. The minimum amount of cover to be inserted in this clause should be at least £5.0 million.

6.2 **Damage to property** is similarly the sole responsibility of the contractor, for which he must also have the necessary insurance to ensure he can meet his obligations. In this instance however he is responsible only for claims arising from his own negligence; there is no provision for any insurance under this contract to cover any damage arising from the negligence of the employer, nor is there provision for the amount of the limit of the indemnity to be inserted (e.g. £50 000 for any one occurrence). The requirement is for 'any damage whatsoever'. Should employers doubt that the contractor has sufficient insurance on a particular project to cover his obligations, this can be remedied by use of the powers under the later clause 6.4 requiring the contractor to provide the necessary evidence.

6.3 **Insurance of the works** against damage arising from the normal perils of fire, flooding, etc, is usually made the responsibility of (A) the contractor for new work, and (B) the employer for alterations. In either case the insurance should be in the joint names of both the contractor and the employer and should cover the full value of the works including landscape architects' fees, unfixed materials, removal of debris and reinstatement and current costs. Any consequential loss to the employer, and to the contractor's temporary buildings and plant are excluded. Where for new works the landscape contractor already has an 'All Risks' policy, the Employer's interest should be endorsed on it accordingly. In the case of *William Tompkinson v. St Michael in the Hamlet* in 1991 in Liverpool, the Judge held that clause 6.3 was concerned only with violent and unusual perils and that as a heavy downfall of rain did not have these characteristics the contractor was responsible for damage by rain to the property of the employer elsewhere on site.

6.4 The clause on **evidence of insurance** provides for the production of such evidence as the employer may reasonably require 'that the necessary insurances have been taken out and are in force'. Although not specifically required, a formal 'Certificate of Insurance' is normally the best way of dealing with this.

6.5 **Malicious damage**, both before and after practical completion, is an increasing problem on landscape projects. Clause 6.5A makes the landscape contractor entirely responsible for making good, at his own expense, all damage arising from theft and malicious damage prior to practical completion. On sites where the risk of such damage to the work is considered to be abnormally high (and the nature of landscape work makes this more common than on building contracts), it may be in the employer's interest to include a provisional sum in the tender conditions to cover the anticipated cost of making good such damage. This is provided for in clause 6.5B. While the contractor is still required to take all steps as are reasonable to minimize the risks of such damage in the event of it occurring, he is then paid no more and no less than the actual cost of the necessary reinstatement without any increase in the total contract sum.

7 Determination

The employment of the contractor (not the contract – an entirely different matter) can be terminated by either the employer or the contractor under the provisions of clauses 7.1 and 7.2. Notwithstanding the apparent simplicity of these two clauses, meticulous and informed legal advice is essential in the drafting and delivery of the necessary notices if the required determination is to be properly achieved and not repudiated.

7.1 **Determination by the employer** can only be made for one of two reasons: either if the contractor has failed 'to proceed diligently' or has wholly suspended the carrying out of the works, or in the event of his having failed financially. In the first case the employer has to prove that the contractor has not merely 'postponed' the work; in the second case it is a matter of fact. In both cases the employer (not the landscape architect) has to 'serve' the notice and the contractor then has to give up the site immediately.

7.2 **Determination by the contractor** can be effected for one of four reasons.

1. If the employer is more than 14 days late in making progress payments.
2. If the employer obstructs the carrying out of the works.
3. If the employer suspends the work for over a month.
4. If the employer fails financially.

Should any one of these occur then the employer has 7 days from receipt of the notice in which to put matters right.

8 The Supplementary Memorandum

This sets out in detail the provisions regarding tax changes, value added tax and the statutory tax deduction scheme which are incorporated by references in clauses 4.5, 5.2 and 5.3.

5.5 Use of the Form in Scotland

The JCLI form was noted as 'not for use in Scotland', it having been drafted in accordance with the provisions of English and not Scots law. The primary difference is that effecting the legally binding nature of the offer and its acceptance. The JCLI form can now be used in Scotland, the line 'subject to the proper law of this contract being English law' having been inserted in the 1st Recital. The parties to any contract can choose the law to apply to it, e.g. civil engineering work in the Middle East is often carried out subject to English law.

It is also necessary to distinguish the law of the place where the work is to be performed, e.g. for the contractor to comply with the statutory obligations, building regulations, and the law of the place where the arbitration referred to in Article 4 is to take place, although this could well follow Scots procedures if required without causing any difficulty.

Similar provisions may apply to work carried out under the laws applicable to Northern Ireland.

5.6 The JCT Intermediate Form IFC84

The JCLI Agreement for Landscape Works was first issued in April 1978 at which time the 1968 JCT Minor Works Form on which it was based had no provision for 'Plant failures' or 'Malicious damage', neither did it cover:

2.6 partial possession;
3.6 the valuation of variations using tender rates;
3.8 p.c. sums and objections to nominations;

3.9 direct loss and expense if progress is disturbed;
4.6 fluctuations;

and detailed provisions were therefore incorporated to cover these aspects.

While suitable for the majority of landscape contracts, instances have occurred of local authorities expressing reluctance to use the Form on landscape contracts in excess of £75 000.

The JCT has now published an 'Intermediate' Form (IFC84) for use on all contracts both private and public sector, local authority, with or without quantities, prepared in accordance with the Standard Method of Measurement (SMM7). While it still does not cover plant failures and malicious damage, the remaining aspects previously omitted are now provided for.

For those larger landscape contracts where the detailed provisions of the JCT Intermediate Form may be appropriate the following supplementary clauses (which are detailed in Appendix D) may be incorporated:

Partial possession by the employer – use IFC84 clause 2.11 (note this is optional in IFC84).
Failures of plants – use JCLI clauses 2.7A and 2.7B, with the last three lines of 4.3.
Malicious damage – use JCLI clause 6.5A.

As with JCT 80, obligatory forms of tender and subcontract are required to be used for 'named' subcontractors (Forms NAM/T and NAM/SC). In this instance, however, the NAM/SC subcontract conditions are incorporated by reference and are not required to be signed as part of the subcontract document.

These two standard forms of lump sum contract, the JCLI generally and the Intermediate Form with the Landscape Supplementary Clauses for large contracts cover all landscape requirements for lump sum contracts, i.e. new work where the scope of the Works is known at the outset and time is available to prepare full drawings, specification and if necessary Bills of Quantities so that tenders can be obtained on a comparable basis and the actual cost of the work is ascertained at the outset.

There are however occasions when the other lump sum forms of contract used for building are incorrectly insisted on by clients, their quantity surveyors or other professional advisors. These are as follows.

5.7 The JCT Form for Minor Building Works (MW80)

This is virtually identical to the JCLI Form but without the additional landscape clauses, and its use for landscape contracts cannot therefore be recommended. The JCT in 1990 also produced a standard form of tender and conditions of contract for jobbing work of a value of up to £10 000 for repairs and minor alterations of a duration of one month or less, and as such is unlikely to be appropriate for use on landscape contracts.

5.8 The JCT Form for Building Works (JCT 80)

First published in 1980 and revised at infrequent intervals subsequently, the 1980 Edition is now a large and complex document suitable only for the multi-million-pound building contract with complex specialist subcontractors. It is not therefore suitable for the relatively simple landscape contracts, even the larger ones. Landscape works are however often incorporated as a subcontract within this building contract and these are discussed separately in the next chapter.

5.9 The general conditions of Government Contracts for Building and Civil Engineering (GC Wks 1&2 Edn 3 1989)

The demise of PSA and autonomy given to government departments and users of buildings such as the Royal Palaces, Museums and Galleries has seen a decline in the use of this lump sum form and its smaller version GC Wks 2 previously obligatory for all government (but not DSS and DoE) work.

The latest edition published in 1989, drafted unilaterally by the government like its predecessors, is primarily for use with quantities although adaptations for other uses are available. It runs to 38 pages of conditions with no Recitals and Articles but has a standard form of tender which, when accepted by the employer, forms the contract. There is no appendix but the same information is given in an abstract of particulars. The 70 conditions are grouped under eight headings similar, but not identical, to the JCLI Form.

It contains several unique features such as stage distinct from periodic payments, and Project Manager with powers delegated to the Clerk of Works. Progress meetings are covered in some detail and grounds for an extension of the contract period are now much more generous although bad weather is no longer recognized as grounds for an extension. Acceleration of the Works by the Employer is now possible and the contract accepts that there may be different Defects Liability Periods, particularly important in the case of landscape works.

The form is clearly written and set out. It is suitable for both building and civil engineering lump sum contracts but like JCT 80 is really drafted for the larger building contracts and is therefore primarily unsuitable for landscape projects.

5.10 The Association of Consulting Architects (ACA 84)

This lump sum contract was first produced in 1982 as a simpler alternative to JCT 80. It provided for with and without quantities projects, in the public or private sectors, fixed price or fluctuations, with detail drawings provided by the architect or the contractor, with or without a list of grounds for an extension of the contract period. Even when revised in 1984 and adapted for use with the British Property Federation system it became almost as complex as the JCT 80 Form it was meant to replace.

Certainly without any supplementary landscape clauses it is unsuitable for the lump sum landscape project, although no doubt in the hands of a competent landscape architect and a competent landscape contractor it could be used successfully. Its innovatory nature and phraseology, although defined, do require different procedures and there is currently no supporting case law to resolve any hidden ambiguities.

In addition to provision for instructions for the acceleration of the works, perhaps the most radical is the addition of 'adjudication' to the alternatives of arbitration or litigation for the resolution of disputes between the Employer or his architect and the contractor. With the current attention being given to ADR (Alternative Dispute Resolution) it would be interesting to know if this has proved successful in relation to the building or landscape industry.

The advent of the JCT Intermediate Form in 1984 (IFC84) has largely eliminated the reasons for the preparation of the ACA form and the existence of the supplementary landscape clauses for use with IFC84 results in its use and not the ACA Form being the recommended option.

5.11 The Architects' and Surveyors' Institute Forms

Formerly two organizations, the Faculty of Architects and Surveyors and the Construction Surveyors' Institute, their amalgamation in 1989 has resulted in the three previous FAS Forms of lump sum contract now being titled ASI Forms although the text remains unchanged.

The 1980 Minor Works, 1981 Small Works and 1986 Building Contracts, similar to the JCT, have been written primarily for Building Contracts and have no clauses specifically for landscape works such as those contained in JCLI and IFC84 with landscape supplement. For this reason they are not to be recommended.

5.12 The Institute of Civil Engineers (ICE) Form 6th Edn 1991

It is essential to realize that the ICE Form is not a lump sum form whereby the works are carried out for a fixed sum, but a remeasurement contract in which the strategic objective is clear, e.g. to build a fixed length of motorway or a bridge across an estuary, but the contractor is paid for the actual quantity of work executed, usually remeasured on a monthly basis valued in accordance with the rates in his tender submitted by the contractor in competition with others.

Although clearly suited to the scale and the unforeseeable and imprecise nature of civil engineering works, which make remeasurement inevitable, it is often used for the larger landscape contract where the large amount of groundwater, movement of earth, roads and footpaths over larger distances make its use appropriate. It also gives the resident engineer and the professional consultants engaged by the employers far wider powers, in respect of the programme sequence of the works and their construction, than is allowed under the JCT and other building contracts.

Other significant differences in the ICE Form, which is 37 pages long with 72 clauses, include the responsibility for damage to roads by the contractor's traffic, interest due on certificates not honoured by the due date, instructions to be issued to accelerate the progress of the works at no extra cost to the employer, and provision for the contractor to give all necessary notices due under the Public Utilities Street Works Act. The Employer's representative is referred to throughout as the Engineer.

There is also a Minor Works Edition first produced in 1988 for contracts of a lesser value than £1 000 000 and a duration of less than six months.

5.13 The New Engineering Contract (NEC)

At the same time as The Institution of Civil Engineers published the sixth edition of the ICE Conditions of Contract in 1991 they also published a draft 'New Engineering Contract'.

Said to be as applicable to building contracts as it was to civil engineering works it introduced an entirely different approach to that in the JCT Forms and involves a Project Manager, who is required to collaborate with the Contractor towards the satisfactory completion of the project but is expected to act solely in the interest of the Employer. It makes no mention of the involvement of an Architect, Engineer or Quantity Surveyor.

The Form of Contract is said to be equally suitable for all lump sum, cost plus or design and build contracts dependent on the clauses selected, and is published as ten separate booklets with a Form of Tender and Schedule of Contract Data in two parts, the first to be completed by the Employer and the second by the Contractor. This is attached to nine groups of Core Clauses applicable to all contracts and one of six sets of Main Option Clauses dependent on the type of procurement required (lump sum, cost plus or design and build etc.), together with such Secondary Option Clauses grouped under 13 headings covering fluctuations, damages for non-completion, performance bonds, contractor's design liability, etc. to be incorporated as and when required.

There is no provision in the draft for nominating subcontractors but it is claimed the conditions are as suitable for subcontractors as they are for main contractors. After having tried the form out in practice the ICE hopes to be able to incorporate any refinements found to be necessary prior to publishing it in its final form.

5.14 Design and Build

There is no doubt that for the smaller and simple project Design/Build has much to commend it; the designer and contractor are one and the same and if anything goes wrong there is only one person to blame. On the other hand the client has forfeited the opportunity to choose the designer, any other consultants and usually any other specialist subcontractors. Time for completion and the cost are certain provided that no changes of mind and variations arise. This type of contract can only succeed if the work is fully detailed and specified at the outset (before the contract is entered into). The client can then expect the quality he believes he is being offered and no hidden cuts will be made during the later stages of the project when the contractor finds his costs are exceeding the tender sum on which his contract is based.

The JCT Standard Form with Contractors Design (WCD 81), intended for building works, deals with the contractor's design responsibilities, planning consents and the valuation of variations and interim certificates, and is recommended for the larger landscape contract procured in this way. Many smaller contracts, however, are entered into on the basis of an acceptance of a landscape contractor's quotation with or without drawings and specification, leaving the detailed matters such as those outlined above unresolved and problems arising subsequently may lead directly to the courts.

Despite the need for the Design and Build contractor to specify in detail what he is providing it is essential that he still remains responsible for the adequacy of its performance if the legal concept of *caveat emptor* is not to be held to apply. While not suggesting that a client is entitled to a first-class product at a very low price the contractor must still remain liable for the adequacy of that which he has quoted for and not be allowed to escape his responsibilities notwithstanding the technical detail he has provided. If the plants subsequently die the responsibility for choice and suitability of species, aspect and adequacy of subsoil and topsoil must remain with the Design/Build Contractor.

5.15 Management Contracts

Prime Cost and Management Contracts are clearly the most suitable type of contract to adopt if early completion is essential and the full scope of the work is unknown at the outset. The quality of the work is ensured by inviting quotations only from contractors and subcontractors of proven ability who are paid no more and no less than the actual cost of the Work executed plus a previously agreed profit element. The overriding disadvantage of course, particularly if the scope is unknown, is that of the uncertainty of the Final Cost and the need for cost-cutting measures, usually affecting the quality of the finished appearance, if it appears during the later stages of the work that the previously agreed target cost is likely to be exceeded.

The 1992 Edition of the JCT Prime Cost Form of Building Contract, issued in a section headed 'format', is equally suitable for landscape projects large and small where an immediate start is required, the main contractor being responsible for all of the work even if he is instructed to place some of it in the hands of specialist subcontractors or suppliers. The Employer may deduct the cost of work improperly executed and excessive use of labour or materials. The main disadvantage is the loss of control over the speed of execution and date for completion, although this can be partially overcome by the use of stage instead of periodic payments or by limiting their frequency.

Management Contracts, the system used for all garden festivals, operate on a similar basis except that the main contractor is precluded from carrying out any of the works himself, all of which he must subcontract as directed, except for providing general attendance himself and remaining responsible for their performance, for which he is paid a lump sum or percentage fee.

Construction Management Contracts are similar to Management Contracts except that all the subcontracts are entered into directly with the Employer and not the Management Contractor who is engaged on similar terms to all the other consultants chosen by the client.

6

Landscape subcontracts

This chapter covers only the general principles of subcontract documentation, highlighting those special characteristics affecting landscape subcontracts. (For detailed consideration of building contracts and subcontracts see *The Standard Form of Building Contract* by John Parris, and *Building Subcontract Forms* by Dennis F. Turner.)

If landscape projects can be carried out under the direction of a landscape architect as a direct contract such as that envisaged by the JCLI conditions between the employer and the landscape contractor this is sometimes in the best interests of all concerned.

6.1 Direct subcontracts – consecutive

Clearly it is simpler if one contractor waits until the other has finished and left the site before starting his own work. This is often possible on projects such as those connected with industry, civil engineering and hospital work, when not only can the landscape work be delayed until the main contractor has cleared up and left the site, but it can also be delayed until the planting can be carried out at a time of year best suited to the establishment of strong and healthy plants.

On landscape work in connection with housing and schools, however, it is essential to have all planting complete and well established before the site is handed over to the future occupants if the plants as well as the buildings are to be treated with the respect to which they are entitled, and the risks of malicious damage minimized.

6.2 Direct subcontracts – concurrent

It is not always possible to execute projects consecutively, and many (such as those connected with buildings) are carried out concurrently with the work of another contractor on the same site at the same time.

'Artists and tradesmen' would seem to be an odd description for a landscape contractor but that was the side heading of the 'Epstein' clause introduced into the earlier forms of the JCT contracts for building work. Under clause 29 of the current form, the employer may, provided that it is described in the tender and contract documents, arrange for others (such as landscape contractors) to enter the main contractor's site and carry out landscape work concurrent with that of the main contractor. Such a division of responsibility must inevitably run great risks of possible dispute on all but the simplest of projects, each side blaming the other – let alone the employer or landscape architect – when something goes wrong. Even when the work is to be carried out by the employer's own staff, e.g. a parks department, great care must be taken to ensure that the main contractor is explicitly informed in the tender documents of the full details of the work involved, if subsequent disputes, claims and arguments are to be avoided.

6.3 Domestic subcontracts – concurrent

By far the simplest procedure, therefore, is to arrange for the landscape work to be carried out by the main contractor using his own staff or subcontractor of his own choosing. This however (unless included as a provisional sum) requires all the decisions with regard to planting – including the preparation of detail drawings, planting layouts, schedules and specification – to have been prepared and included in the tender documents. If Bills of Quantities are involved the work must be properly measured and included in the Bills. Clause 19 of the JCT Standard Form of Building Contract then permits the main contractor either to carry out the work himself or to subcontract the work to a firm entirely of his own choosing, possibly one with whom he has already established a good working relationship in the past. Should he decide to subcontract the work this is then subject to the right of reasonable objection by the employer, but the main contractor still accepts full responsibility for its satisfactory completion; he has to arrange for it to be carried out at a time to suit his own convenience and programme, compatible with the needs of the plants for seasonal lifting and their subsequent establishment.

It is important to note that by including the work within the main contract in this way, the employer can withold his Certificate of Practical Completion should he wish, until the landscape works are also complete.

The advantages of having a standard form of subcontract for use under these circumstances are obvious, each party knowing that it is compatible with the main Form, in time becoming familiar and experienced with its usage, knowing that all aspects and their own requirements are fully covered, and that any special requirements are clearly identified.

The JCLI has now prepared a standard form of subcontract (Appendix B) for use with the JCLI Standard Form of Agreement for Landscape Works (Appendix C). This form incorporates, for the information of subcontract tenderers, the conditions of the Main contract and on the reverse the Form of Tender for the subcontractor works on which the subcontractor enters his price; to this are attached the Conditions drafted to mirror the obligations of the Employer and Main Contractor into the subcontract.

Such a Form also exists for use with the JCT 80 Standard Form of Building Contract. The standard form of subcontract, for Domestic Subcontractors appointed under clauses 19.2 and 19.3 of the 1980 JCT Standard Form of Building Contract, is suitable for use in all editions: Local Authorities or Private, With or Without Quantities. Although prepared by the Building Employers Confederation (BEC), the Federation of Associations of Specialists and Subcontractors, and the Committee of Specialist Engineering Contractors, its use ensures that the obligations of the main contractor are effectively transferred to the subcontractor and are therefore equally appropriate for landscape subcontractors on building contracts.

The actual form is published in two parts, the first being the Articles of Agreement comprising the four Recitals, identifying the scope of the subcontract works, the main contract works, and the right of inspection of the main contract conditions. The three Articles cover the main and subcontract provisions, obligations and conditions, the subcontract sum (either lump sum or re-measured), an arbitration clause providing for an Arbitrator appointed by the president of the RICS, and for the law of subcontract sum to be the law of England unless otherwise stated. The form provides also for the subcontract sum to be entered into the Form and for it to be signed by the main contractor and subcontractor, under hand or seal as required.

The second part comprises the subcontract conditions which can be incorporated by reference under Article 1.3. The Conditions run to 37 clauses over 44 pages which, although they adopt decimal numbering, are neither section-headed nor follow the same sequence as JCT 80. They do, however, deal exhaustively with all aspects of subcontract work likely to arise on the larger building and specialist engineering subcontract under the headings listed in Table 6.1.

Table 6.1 Standard form of subcontract for use with JCT 80 Standard Form of Building Contract – main clause headings

1. Interpretation.
2. Subcontract documents.
3. Subcontract sum additions.
4. Directions for subcontract works.
5. Subcontract liability.
6. Injury to persons/property.
7. Subcontract insurance.
8. Clause 22 perils.
9. Insurance policies.
10. Responsibility for subcontract plant.
11. Subcontract date for completion.
12. Failure to complete.
13. Direct loss and expense.
14. Subcontract Practical Completion.
15. Subcontract price.
16. Valuation of variations.
17. Valuation of subcontract works.
18. Bills of Quantities – SMM.

19. Value added tax.
20. Finance (No.2) Act 1975.
21. Payments of subcontractor.
22. Benefits under main contract.
23. Right to set off.
24. Adjudicating subcontract claims.
25. Right of access.
26. Assignment of subletting.
27. Attendance.
28. Interference with property.
29. Determination of subcontract by main contractor.
30. Determination of subcontract by subcontractor.
31. Determination of main contract.
32. Fair wages.
33. Loss and expense by strikes.
34. Fluctuations method.
35. Contractors' levies and taxes.
36. Fluctuations in cost.
37. Formula adjustment.

The BEC also published in August 1984, as Amendment No. 1, an additional clause 21.4.5 relating to the ownership of unfixed subcontract materials and goods on site which have been included in an Interim Certificate so that they become the property of the Employer when the Employer has honoured the certificate and paid the main contractor. This effectively overcomes the Romalpa problems of the Dawber Williams case relating to roofing tiles, although this view is not universally held.

6.4 Listed subcontractors

The particular nature of landscape works, with the problems of plant supplies and the difficulties in the precise specification of workmanship and materials, often makes it advisable to restrict the main contractor's choice of subcontractors for the landscape work to only those of proven experience and ability. To ensure some control over the choice of suitable landscape subcontractors, employers and their landscape architects often specify – as suggested in section 5.4.3 – a list of three or four firms of an acceptable standard, restricting the choice of main contractor to the firms named from whom he is to obtain tenders. At the same time he is allowed, at his sole discretion, to choose the one to carry out the work, then accepting full responsibility for their satisfactory performance, making his own arrangements for payment, and deciding when the subcontract work is to be carried out.

This is also covered by clause 19 of the JCT Form of Building Contract. Although it provides for either party with the consent of the other (which shall not be unreasonably withheld) to add additional firms to the list at any time, the specification should include a clause requiring the main contractor to request approval to any addition or substitution 10 days prior to the tender date. A suitable form of enquiry to take up references on subcontractors put forward by main contractors is shown in Figure 6.1. It is important to ensure that when submitting their tender, contractors should be asked to name which of the listed subcontractors has been chosen and included.

A subcontractor selected by the main contractor in this way remains a domestic subcontractor, so that the main contractor has no right to an extension

PRIVATE & CONFIDENTIAL

Dear Sirs,

We have been given your name by the above in connection with their request to be considered for work to be carried out under our direction.

We would be very grateful if you would complete the attached copy of the following questionnaire and return it to us in the envelope provided as soon as possible.

(Please indicate by ✓)	Below Av.	Average	Above Av.
1. Office and site organization			
2. Standard of workmanship			
3. Observance of completion dates			
4. Availability of labour and co-ordination of sub contractors and suppliers			
5. Willingness to make good defects			
6. Absence of unreasonable claims and prompt settlement of accounts			

7. What would you consider to be the maximum value of any one contract they could undertake? £ ....

8. Would you use them again? YES/NO

9. Any other comments:

Yours faithfully,

Fig. 6.1 Example letter of enquiry to take up references on subcontractors.

of time for any delay on the part of the subcontractor and also accepts full responsibility for the performance of the subcontractor he has selected. He accepts sole responsibility for conduct and execution on site, having also to sort out any delayed starting dates, sequence of trades, and disputes with others employed on the works at the same time. This is a most valuable provision in building contracts, and may become the standard way of selecting subcontractors for work on building contracts in the future. It has been suggested that it may not produce the lowest tenders, but as there is every incentive for the main contractor to select not the lowest but the most reliable subcontractors, this would appear to be unlikely, and the knowledge that the lowest tenderer has included the most reliable and suitable subcontractors at the same time accepting responsibility for their proper performance would suggest its increasingly widespread adoption.

6.5 Nominated subcontractors on lump sum building contracts

On those occasions when for one reason or another it has not been possible to measure and accurately describe the landscape work in the tender documents, or it is necessary to restrict the work to a **single** landscape subcontractor chosen by the employer or his landscape architect, for one of the three reasons said by Banwell to justify nomination, i.e.

1. when special techniques are required, or
2. when early ordering of materials is necessary, or
3. when workmanship of a particular quality is essential

(all of which apply to landscape work) then under clause 19 of the JCT Standard Form, this requires the inclusion of a prime cost sum in the tender documents and for the work to be carried out as a nominated subcontract under clause 35.

As the procedures to be followed when a landscape subcontractor is nominated under clause 35 have been devised so that they may be applied to all building nominated subcontracts, e.g. heating and electrical engineering, piling, fixing external cladding and the like, much of the detail such as that relating to shop drawings is inapplicable to landscape work, but even had the procedures not been made obligatory, landscape subcontractors on building contracts would still have needed the same information to be provided in their tender documents.

JCT 80 Clause 35

The effect of this clause is that, in return for the right to restrict the work to a single landscape subcontractor of his own choosing, the employer loses his right to liquidated damages if the main contract completion is delayed by the landscape subcontractor, and the landscape subcontractor is entitled to direct payment if the main contractor does not pay amounts stated as being due to him in Interim Certificates.

There was considerable criticism of the complexity of the subcontract procedures when the revised edition of the standard form was published in 1980 although these were considered essential if the identity of the main contractor and subcontractor was not known to the other at the time of their tender. The forms therefore ensured that any special requirements of the subcontractor were made known to the main contractor, and the periods when the main contractor required the subcontractor to carry out the work were made known to the subcontractor before the subcontract was entered into.

If this had not been done when the nominated subcontractor had submitted his tender, JCT 80 had an 'alternative' procedure to be carried out before the subcontract was entered into.

In answer to the criticism the JCT revised its nominated subcontract procedures in March 1991, combining both methods of nomination into one and issued the following new forms.

NSC/W

The problem of the loss of liquidated and ascertained damages is overcome in clause 35 by the use of a separate agreement between the employer and subcontractor in which they enter into a direct contractual arrangement (NSC/W) where clause 3.4 states that:

The subcontractor shall so perform the Subcontract that the Contractor will not be entitled to an extension of time for completion of the Main Contract works by reason of the Relevant Event in clause 24.4.7 of the Main Contract conditions.

NSC/W also requires the subcontractor to warrant that he has exercised, and will exercise, all reasonable skill and care in the design of any of the work, in the selection of any goods or materials, and in the satisfaction of any performance specification for which he is responsible.

The employer is then entitled to recover from the subcontractor the liquidated and ascertained damages he has lost as a result of having had to grant an extension to the contract period of the main contractor by reason of the delay caused by the nominated landscape subcontractor. For this reason – so that the nominated subcontractor is aware of the amount of liquidated damages and the contract period in the main contract – the use of a standard Form of Tender containing this information (NSC/T) is also made obligatory.

NSC/T

This standard Form of Tender, the use of which is obligatory, ensures that all the information required both by the main contractor and the subcontractor is known to each other before they enter into their subcontract.

The key sections of this document are in the two pages (2 and 7) (Figures 6.2 and 6.3) completed in turn by the landscape architect, the landscape subcontractor, and the main contractor before the subcontract is entered into. Figure 6.2 demonstrates a worked example.

The landscape architect, before sending the Form of Tender to the subcontractor(s), inserts the period during which he estimates the site will be available for the landscape works to be carried out on site.

Subsequently, the landscape subcontractor inserts the period he estimates he will actually need to complete satisfactorily the work for which he is tendering, together with other timetables for ordering material and the notice he requires to commence work. He then returns the Form to the landscape architect.

Finally, the main contractor, before entering into the subcontract, inserts any further details as to the sequence of work he may wish to impose and any revisions to the programme for the execution of the work.

Part 3 – Particular Conditions

Items to be completed by the Contractor and the Sub-Contractor

Notes on completion of these Particular Conditions

[a] Insert the same details as in NSC/T Part 1, pages 2 and 3.
In NSC/T Part 3 the expression 'Contract Administrator' is applicable where the Nomination Instruction on Nomination NSC/N will be issued under a Local Authorities version of the Standard Form of Building Contract and by a person who is not entitled to the use of the name 'Architect' under and in accordance with the Architects (Registration) Acts 1931 to 1969. If so, the expression 'Architect' shall be deemed to have been deleted throughout Tender NSC/T. Where the person who will issue the aforesaid Nomination Instruction is entitled to the use of the name 'Architect' the expression 'Contract Administrator' shall be deemed to have been deleted throughout NSC/T.

[b] The Sub-Contractor's entries in NSC/T Part 2, items 1 to 4 should be considered when agreeing the entries in NSC/T Part 3.

[c] Conditions NSC/C state in clause 2·1: 'The Sub-Contractor shall carry out and complete the Sub-Contract Works in accordance with the agreed programme details in NSC/T Part 3, item 1, and reasonably in accordance with the progress of the Works but subject to receipt of the notice to commence work on site as detailed in NSC/T Part 3, item 1, and to the operation of clauses 2·2 to 2·7.' *(extension of Sub-Contract time).*

[d] The period of notice **must** take account of any period stated for the execution of the Sub-Contract Works off-site prior to commencement on site

[a] Main Contract Works and location:
Restoration Chiswick House
Burlington Lane
London W4

[a] Sub-Contract Works:
Landscape work

Specimen

[b]

1 (1) Period required by the Architect/the Contract Administrator to approve drawings after receipt will be that set out in NSC/T Part 1, item 14.
Not Applicable

[c] (2) The earliest starting date and the latest starting date for the Sub-Contract Works to be carried out on site:
are Commence on site 7 February 1995 (earliest)
and Complete on site 30 March 1995 (latest)

(3) Periods required for:

(i) submission for approval of all necessary Sub-Contractor's drawings etc. *(co-ordination, installation, shop or builders' work or other as appropriate)* NA weeks

(ii) the execution of the Sub-Contract Works

off-site (if any) prior to commencement on site NA weeks

on site weeks

[d] from expiry of **period required for notice to commence work on site** which is 7 weeks

(4) Further details:
Complete 1 courtyard at a time commencing at the North end of the site and working Southwards

Fig. 6.2 Worked example from Tender NSC/T Part 3 page 2. (Reproduced with permission of RIBA Publications Ltd.)

Notes

[h1] For Agreement executed under hand and NOT as a deed.

[h2] For Agreement executed as a deed under the law of England and Wales by a company or other body corporate: insert the name of the party mentioned and identified on page 1 and then use *either* [h3] and [h4] *or* [h5]. If the party is an *individual* see note [h6].

[h3] For use if the party is using its common seal, which should be affixed under the party's name.

[h4] For use of the party's officers authorised to affix its common seal.

[h5] For use if the party is a company registered under the Companies Acts which is not using a common seal: insert the names of the two officers by whom the company is acting *who MUST be either a director and the company secretary or two directors*, and insert their signatures with 'Director' or 'Secretary' as appropriate. *This method of execution is NOT valid for local authorities or certain other bodies incorporated by Act of Parliament or by charter if exempted under s.718(2) of the Companies Act 1985.*

[h6] If executed as a deed by an *individual:* insert the name at [h2], delete the words at [h3], substitute 'whose signature is here subscribed' and insert the individual's signature. The individual MUST sign in the presence of a witness who attests the signature. Insert at [h4] the signature and name of the witness. Sealing by an individual is not required.

Other attestation clauses are required under the law of Scotland.

[h1] **AS WITNESS THE HANDS OF THE PARTIES HERETO**

[h1] Signed by or on behalf of the Employer____________________

in the presence of:

[h1] Signed by or on behalf of the Sub-Contractor____________________

in the presence of:

Complete under hand (above) or as a deed (below) as applicable: see NSC/T Part 1, page 2.

[h2] **EXECUTED AS A DEED BY THE EMPLOYER**
hereinbefore mentioned namely____________________

[h3] by affixing hereto its common seal

[h4] in the presence of:

* OR

[h5] acting by a director and its secretary* / two directors* whose signatures are here subscribed:
namely____________________

[Signature]____________________*DIRECTOR*

and____________________

[Signature]____________________*SECRETARY* /DIRECTOR**

[h2] **AND AS A DEED BY THE SUB-CONTRACTOR**
hereinbefore mentioned namely____________________

[h3] by affixing hereto its common seal

[h4] in the presence of:

* OR

[h5] acting by a director and its secretary* / two directors* whose signatures are here subscribed:
namely____________________

[Signature]____________________*DIRECTOR*

and____________________

[Signature]____________________*SECRETARY* /DIRECTOR**

* *Delete as appropriate*

Page 7

Fig. 6.3 Tender NSC/T Part 3 page 7. (Reproduced with permission of RIBA Publications Ltd.)

The actual procedure is carried out in successive stages as follows.

1. The landscape architect invites tenders as NSC/T Part 1 by sending it with the other tender documents including NSC/W to the subcontractor concerned, who returns the forms with Part 2 completed.
2. The landscape architect arranges for the Employer to countersign the selected subcontractor's tender and also the Employer subcontractor agreement NSC/W and then sends all the signed documents to the main contractor with copies to the selected subcontractor.
3. The main contractor agrees with the subcontractor concerned and completes Part 3 regarding the programme and any other outstanding details before entering into the subcontract agreement NSC/A which incorporates by reference the standard subcontract conditions NSC/C, and which therefore remains 'on the shelf'.
4. The main contractor and subcontractor have 10 days from receipt of the landscape architect's nomination Form NSC/N to complete Part 3 and if they are unable to do so within this period must refer the matter to the landscape architect.
5. Both main contractor and subcontractor are entitled to refuse the nomination or withdraw their tender when the identity of the other becomes known to them, and if they are unable to resolve any differences the landscape architect must:
   (a) if he does not consider that the matters in dispute are sufficiently fundamental to justify non-compliance either extend the period to enable the matters in dispute to be resolved or instruct the contractor to accept the nomination.
   (b) if he considers that the matters justify non-compliance either revise his instructions to enable the matters to be resolved, omit the work or nominate another subcontractor.

The JCT Standard Form of Nominated Subcontract Agreement and Conditions NSC/A and NSC/C

The subcontract documents now come in two parts. The Agreement (NSC/A), a seven-page form to be completed by main contractors and subcontractors subsequent to their nomination for each individual subcontract, and the actual subcontract conditions (NSC/C) incorporated by reference into the Agreement thereby remaining 'on the shelf'. They have been specifically designed to reflect in every respect the provisions of the main contract, particularly those dealing with insurance, disturbance to progress, and tax certificates. No such JCT subcontract previously existed, and its use has caused few (if any) problems to date.

The subcontract conditions include three contents pages setting out headings of the 34 main clauses, and runs to 42 pages. It is not intended that separate NSC4 subcontracts will be signed for each nomination, as they are incorporated in NSC/A by reference. For full details the JCT guidance notes should be referred to, but in general terms the same provisions for nominated subcontractors, as were included in the 1963 JCT Edition, are unchanged, i.e. those affecting direct payment, early release of retention, responsibility for attendance, making good defects, extensions to the contract period, adjudication, and determination.

The provisions for the right of set-off by the main contractor are set out in clauses 23 and 24; they deal both with the case when it has been agreed with the contractor and subcontractor, and when it has been awarded by the court in litigation or arbitration. The payment of the costs of both an arbitrator and an adjudicator are also covered.

The subcontract also provides, under clause 29, for the architect to determine the subcontract if he considers the subcontractor to be in default, if the subcontractor becomes insolvent, or if the subcontractor himself has validly determined the subcontract. The architect then is obliged to make a further nomination and, if this is because of a valid determination by the subcontractor, then any extra cost to the employer arising from the new nomination has to be borne by the main contractor.

6.6 Subcontracts executed under the ICE Form of Contract

The standard form of subcontract prepared for use with the ICE Form is relatively simple, with the conditions of the main contract incorporated by reference. Nineteen short clauses on seven pages are followed by five schedules detailing:

1. the particulars of the main contract;
2. the subcontract documents and scope of the works;
3. the price, retention, and period for completion;
4. facilities provided by the contractor;
5. insurances.

The subcontract clause headings are listed in Table 6.2.

Table 6.2 Standard form of subcontract for use with ICE Standard Form – main clause headings

1.	Definitions.	11.	Property, materials, and plant.
2.	General.	12.	Indemnities.
3.	Main contract.	13.	Insurances.
4.	Contractor's facilities.	14.	Maintenance and defects.
5.	Site working access.	15.	Payment.
6.	Commencement and completion.	16.	Determination of main contract.
7.	Instructions and decisions.	17.	Subcontractors' default.
8.	Variations.	18.	Disputes.
9.	Valuation of variations.	19.	Value added tax.
10.	Notices and claims.		

7

Plant supplies

The success of any landscape project is dependent on the satisfactory synthesis of:

1. the choice of the appropriate plant;
2. the cultivation of sufficient quantities of healthy stock by the nurseryman;
3. its successful lifting, packaging and delivery to site;
4. the proper handling of the plants after they have been delivered.

The failure of any one aspect will inevitably result in the death of the plant and possibly the failure of the project even before the date of practical completion.

Choosing the appropriate plant requires a combination of an appreciation not only of the physical aspects of the plants such as the height, spread and seasonal changes of colour of leaves, flowers, bark and fruit, but also of the microclimatic conditions, the preferred type and pH value of the soil, the requirements for sun and shade, and toleration of wind, frost, salt, draught and urban pollutants, on which its survival depends.

No doubt all this information will be shortly available on instant access from a computerized data bank. At present one must depend on memory and a variety of sometimes conflicting technical references, never fully comprehensive, and often having to resort to experience and skilled judgement, usually known as plantsmanship, an attribute sometimes thought to be possessed by a rapidly declining number of people.

7.1 Choosing the appropriate plant – using the JCLI plant lists

To ensure that, having selected the appropriate plant, it is available in sufficient quantities when required, it is necessary to rationalize cultivars – particularly those with a genus where previously many almost identical cultivars, e.g. rhododendron, azalea and roses, have been on offer by nurserymen.

Landscape designers and managers, both in the local authority and in the private sector, need a list from which they can confidently select plants in the

Fig. 7.1 JLCPS/CPSE Herbaceous plant list.

Two lists are included in this publication –

A. Exotic herbaceous plants (pages 1 - 19)

B. British Native herbaceous plants (pages 20 - 28)

The objectives of both lists are:

1. To promote the more widespread use of herbaceous plants by means of two definitive, informative lists which summarise useful information not readily available elsewhere.

2. In the Exotic List to rationalise cultivars, hybrids or strains as far as is possible excepting that in the case of those Genera marked thus * a selection of the many and often ephemeral cultivars etc. is advised in consultation with reputable suppliers.

3. To improve the future availability of these plants in Britain particularly of native species considered to be increasingly in demand during the next few years.

(A) EXOTIC HERBACEOUS PLANTS

EXPLANATORY NOTES

1. This list includes 'typical' herbaceous plants, non-woody ground cover and the more vigorous rock plants; clump-forming bulbous species, ornamental grasses and sedges, ferns and aquatics. It excludes annuals, biennials and most bulbous species and most herbs. The majority of plants are introduced species and cultivars or hybrids (i.e. **exotic** or non-native). Some native aquatic plants are included.

2. The alphabetical lists are printed in bold and in normal type to distinguish priority and important plants from the other selections.

3. The column heading used in conjunction with the category markings as denoted should be used as guidelines only. In some cases information is not available for every plant species listed. This is particularly the case with the last column that refers to the use of the herbicide Lenacil.* It should be emphasised that any herbicides, including Lenacil should be used with caution on herbaceous plants but as Lenacil probably has the greatest tolerance in this respect its use can be advised with care as indicated. Simazine may also be used with great care and discrimination on some herbaceous plants.

4. Where it is considered inadvisable or misleading to specify or recommend named cultivars or hybrids of genera such as Delphinium and Lupin marked thus *, a selection from a current reputable supplier's list is recommended.

*Information kindly supplied by the Weed Research Organization.

Fig. 7.1 (*Contd*)

knowledge that they are likely to be generally available in quantity from most of the reputable nurserymen throughout the country. Garden centre operators need to know which species to stock in quantity to meet current and future demands. Nurserymen require long-term guidance on the plants they will be required to supply two or three years hence. Landscape and horticultural students need to learn the characteristics of a basic list of plants, to be supplemented in later years as the students' range of experience and skill grows.

To meet these demands JLCPS and JCLI have prepared two lists of plants – the herbaceous list (Figure 7.1), and the tree and shrub list (Figure 7.2) –

JLCPS HERBACEOUS

(A) EXOTIC HERBACEOUS LIST

heavy type indicates priority plants

plant falls clearly into category shown ●

plant may be suitable for category shown ○

Lenacil column ○ Some tolerance

● Tolerant

	Type					Best Growth Conditions						Other factors					
	Rock Plant	Typical herbaceous plant	Fern	Grass, ornamental/sedge	Aquatic	General purpose	Sun	Shade	Dry	Moist	Waterside, Marginal	Winter effect	Ground cover	Long lived	Invasive	Spring or ex-pot planting	Tolerant of Lenacil
ACAENA 'Blue Haze'	●						●		●				●		○	●	
microphylla	●						●		●				●		○	●	
ACANTHUS mollis		●				●	●	○	●	●		○		●	○		○
spinosus		●				●	●	○	●	●		○		●	○		○
ACHILLEA 'Coronation Gold'		●				●	●	○	●			●					○
'Moonshine'		●				●	●	○	●			●					○
ptarmica 'The Pearl'		●				●	●	○	●			○					○
millefolium 'Cerise Queen'							●	○	●				●		○	●	○
ACONITUM x cammarum 'Blue Sceptre'		●				●	●	○	○	●				●			●
(Monkshood) **'Bressingham Spire'**		●				●	●	○	○	●				●			●
ACTAEA spicata		●					○	●		●							
AETHIONEMA 'Warley Rose'	●					●	●		●				○			●	
AGAPANTHUS Headbourne Hybrids		●					●		●							●	○
AJUGA reptans 'Burgundy Glow'							○	●	○	●		●	●			●	
(Bugle) **'Purpurea'**							○	●	○	●		●	●			●	
'Variegata'							○	●	○	●		●	●			●	
ALCHEMILLA mollis		●				●	●	●	●	●			●	●	○		●
ALISMA plantago aquatica (Water Plantain)					●		●				○					●	
ALSTROEMERIA Ligtu Hybrids		●				●	●		●						○	●	
ALYSSUM saxatile citrinum	●						●		●			○	○			●	
ANAPHALIS triplinervis		●				●	●	○	●	●		○					●
ANCHUSA azurea 'Loddon Royalist'		●				●	●		●							●	●
ANDROSACE sarmentosa	●					●	●		●				○			●	
ANEMONE x hybrida (including A. **hupehensis** and A. **japonica**)																	
'Honorine Jobert'		●				●	●	○	○	●				●	○		○
'September Charm'		●				●	●	○	○	●				●	○		○
'White Queen'		●				●	●	○	○	●				●	○		○

Fig. 7.1 (*Contd*)

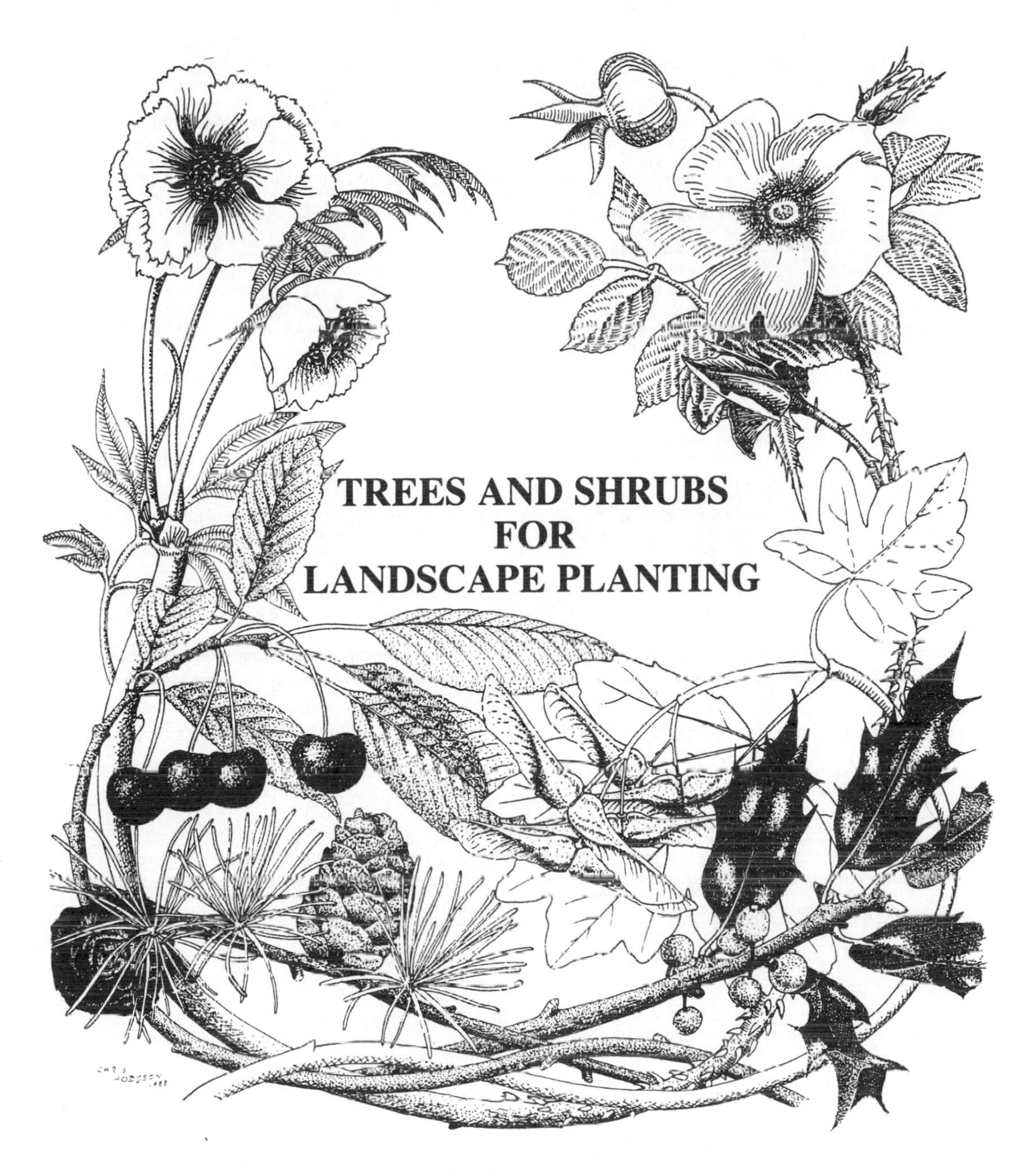

A list of trees, shrubs, conifers and climbing plants recommended by the Joint Council for Landscape Industries.

Fig. 7.2 JCLI tree and shrub list.

NOTES FOR GUIDANCE

Each of the four groups of woody plants in the list is given a separate section: Trees, Shrubs, Climbers and Conifers. Bamboos are included in the shrub section. The information on the plants within these groups is to be found in the columns under four main categories: *Habit, Soils, Aspect* and *Performance,* each of these categories then divides into a number of sub-headings which form the column heads. Due to the different characteristics of the four main groups of woody plants, the same set of columns cannot be used for all and some variations should be noted. The following notes are included to aid understanding of those columns which are not immediately self explanatory:

Columns 1-4: Habit
Size is assessed on a different scale for each of the four main categories, and refers to the fully grown plant. Actual dimensions are not suggested since they will vary according to soil, climate and other environmental factors. It is accepted that difficulties arise with such categories as large shrubs or small trees which could be in either group. Where appropriate cross references are used. In the shrub and conifer groups columns the letter "P" indicates a prostrate habit. In the Conifers section a special column has been included to differentiate between tree and shrub forms.

Columns 5-6: pH Tolerance
Column 5 indicates those calcifuge plants that normally prefer or need lime-free or acidic conditions. Column 6 indicates those plants that will grow quite well over a wide range of soil pH from alkaline to acid. It should be remembered however that whereas almost all the plants that will thrive on calcareous or high pH soils (calcicoles) will usually do well on acidic soils, the reverse is not true.

Column 12: Hardiness
The following three degrees of increasing hardiness shown must be regarded as guidelines only since there can be wide variations of microclimate in the UK as well as extremes of weather conditions over a period of years:

1. May benefit from some protection, or use is advised in warmer sheltered and protected sites.
2. Normally hardy from the midlands, south and westwards and in favoured places further north. (Some protection may be needed in local colder areas).
3. Normally hardy anywhere in the UK.

Note however, that the very fact that these woody plants are included in this list indicates generally a good measure of resistance and a capacity to survive or recover from abnormal winter conditions.

Column: Growth Rates (number of column varies with group)
Growth rates will naturally vary depending on the conditions available to the plant, but assuming normal growth rate under reasonable conditions the three rates are:

F = Fast
M = Moderate
S = Slow

Column: Effective Life (number of column varies with group)
This indicates the normal life expectancy of a woody plant under reasonable cultural conditions accepting that pruning and other remedial techniques may prolong the life of some woody plants, especially trees. These can only be guidelines, and the life expectancy indicated is within the respective category re trees or shrubs etc.

Column: Pest and Disease Susceptibility (number of column varies with group)
Only the most likely pest and diseases are indicated with the appropriate symbol and a note in the 'other factors' column. The most serious disease at this time is 'Fireblight', a *notifiable disease;* and any known or suspected outbreaks must be reported to the Ministry of Agriculture, Fisheries and Food (MAFF) through the regional MAFF office where further information can be obtained on symptoms, control etc.

Fig. 7.2 (*Contd*)

Column: Flower Period (number of column varies with group)

The main flowering period is shown by the numbers of the months. In very favoured or very cold areas some divergence may occur

Column: Flower Colour (number of column varies with group)

The following abbreviations are only approximate guides to the colour group and further references must be sought for the many different shades and hues.

WH	=	white	BL	=	blue
PK	=	pink	YL	=	yellow/cream
RD	=	red	OR	=	orange
VT	=	violet			

Column: Other Factors

Availability factor. This refers to those plants where due to propagation problems or for other reasons, supplies may be limited.

Symbols

A key to the symbols and abbreviations used in the columns is set out below and also provided on each right hand page. A black circle (●) indicates that the plant falls clearly into the category, an open circle (○) that the plant may be suitable for the category

KEY TO SYMBOLS AND ABBREVIATIONS FOR **TREE** SECTION

HABIT	
1 LARGE	●
2 MEDIUM	●
3 SMALL	●
4 EVERGREEN	E
SEMI EVERGREEN	SE
SOILS	
5 DEMANDS ACID SOIL	●
6 WIDE pH TOLERANCE	●
7 PERMANENTLY WET	●
8 HEAVY, MOISTURE RETAINING	●
9 MEDIUM LOAM	●
10 LIGHT, FREE-DRAINING	●
11 INFERTILE SOIL TOLERANT	●
ASPECT	
12 HARDINESS 1, 2, 3	3
13 EXPOSURE: COASTAL (SALT TOLERANT)	●
14 EXPOSURE: INLAND	●
PERFORMANCE	
15 GROWTH RATE:	
SLOW	S
MEDIUM	M
FAST	F
16 EFFECTIVE LIFE:	
SHORT	S
MEDIUM	M
LONG	L
17 PEST/DISEASES:	
HIGHLY SUSCEPTIBLE	●
MODERATELY SUSCEPTIBLE	○
18 TOUGHNESS: SURVIVOR/PIONEER	●
19 SHELTER/SCREEN	●
20 BROAD SPREAD	●
21 NARROW SPREAD	●
22 COLUMNAR	●
23 WEEPING FURNISHED TO THE GROUND	●
24 OVERALL FOLIAGE TEXTURE:	
FINE TEXTURE	F
MEDIUM TEXTURE	M
BOLD TEXTURE	B
25 OVERALL FOLIAGE COLOUR:	
YELLOW	YL
PURPLE	PR
SILVER/GREY	SV
VARIEGATED	VR
26 AUTUMN LEAF COLOUR	L
FRUIT	FR
27 FLOWERING PERIOD	7-10
28 FLOWER COLOUR:	
WHITE	WH
PINK	PK
RED	RD
VIOLET	VT
BLUE	BL
YELLOW/CREAM	YL
ORANGE	OR
29 SPINES OR THORNS	●

KEY TO SYMBOLS AND ABBREVIATIONS FOR **SHRUB** SECTION

HABIT	
1 LARGE	●
2 MEDIUM	●
3 SMALL/PROSTRATE	● P
4 EVERGREEN	E
SEMI EVERGREEN	SE
SOILS	
5 DEMANDS ACID SOIL	●
6 WIDE pH TOLERANCE	●
7 PERMANENTLY WET	●
8 HEAVY, MOISTURE RETAINING	●
9 MEDIUM LOAM	●
10 LIGHT, FREE-DRAINING	●
11 INFERTILE SOIL TOLERANT	●
ASPECT	
12 HARDINESS 1, 2, 3	3
13 EXPOSURE: COASTAL (SALT TOLERANT)	●
14 EXPOSURE: INLAND	●
15 MAXIMUM SUN, S.WALL	●
16 FULL SUN PREFERRED	●
17 SHADE TOLERANT	○
HEAVY OVERHEAD SHADE	●
PERFORMANCE	
18 GROWTH RATE:	
SLOW	S
MEDIUM	M
FAST	F
19 EFFECTIVE LIFE:	
SHORT	S
MEDIUM	M
LONG	L
20 PEST/DISEASES:	
HIGHLY SUSCEPTIBLE	●
MODERATELY SUSCEPTIBLE	○
21 TOUGHNESS: SURVIVOR/PIONEER	●
22 HEDGE	HG
SCREEN	SC
23 GROUND COVER	●
24 FORM:	
SHEAF/VASE SHAPED	SH
MOUND/DOME	MD
25 OVERALL FOLIAGE TEXTURE:	
FINE TEXTURE	F
MEDIUM TEXTURE	M
BOLD TEXTURE	B
26 OVERALL FOLIAGE COLOUR:	
YELLOW	YL
PURPLE	PR
SILVER/GREY	SV
VARIEGATED	VR
27 AUTUMN LEAF COLOUR	L
FRUIT	FR
28 FLOWERING PERIOD	7-10
29 FLOWER COLOUR:	
WHITE	WH
PINK	PK
RED	RD
VIOLET	VT
BLUE	BL
YELLOW/CREAM	YL
ORANGE	OR
30 SPINES OR THORNS	●

KEY TO SYMBOLS AND ABBREVIATIONS FOR **CLIMBER** SECTION

HABIT	
1 LARGE	●
2 MEDIUM	●
3 SMALL	●
4 EVERGREEN	E
SEMI EVERGREEN	SE
SOILS	
5 DEMANDS ACID SOIL	●
6 WIDE pH TOLERANCE	●
7 PERMANENTLY WET	●
8 HEAVY, MOISTURE RETAINING	●
9 MEDIUM LOAM	●
10 LIGHT, FREE-DRAINING	●
11 INFERTILE SOIL TOLERANT	●
ASPECT	
12 HARDINESS 1, 2, 3	3
13 EXPOSURE: COASTAL (SALT TOLERANT)	●
14 EXPOSURE: INLAND	●
15 MAXIMUM SUN, S.WALL	●
16 FULL SUN PREFERRED	●
17 SHADE TOLERANT	●
PERFORMANCE	
18 GROWTH RATE:	
SLOW	S
MEDIUM	M
FAST	F
19 EFFECTIVE LIFE	
SHORT	S
MEDIUM	M
LONG	L
20 PEST/DISEASES:	
HIGHLY SUSCEPTIBLE	●
MODERATELY SUSCEPTIBLE	○
21 TOUGHNESS: SURVIVOR/PIONEER	●
22 SUPPORT METHOD	
SELF CLINGING	CL
SPRAWLING	SP
TWINING	TW
23 GROUND COVER	●
24 OVERALL FOLIAGE TEXTURE:	
FINE TEXTURE	F
MEDIUM TEXTURE	M
BOLD TEXTURE	B
25 OVERALL FOLIAGE COLOUR	
YELLOW	YL
PURPLE	PR
SILVER/GREY	SV
VARIEGATED	VR
26 AUTUMN LEAF COLOUR	L
FRUIT	FR
27 FLOWERING PERIOD	7-10
28 FLOWER COLOUR:	
WHITE	WH
PINK	PK
RED	RD
VIOLET	VT
BLUE	BL
YELLOW/CREAM	YL
ORANGE	OR

KEY TO SYMBOLS AND ABBREVIATIONS FOR **CONIFER** SECTION

HABIT	
1 TREE/SHRUB FORM	TR SH
2 LARGE	●
3 MEDIUM	●
4 SMALL/PROSTRATE	● P
5 DECIDUOUS	D
SOILS	
6 DEMANDS ACID SOIL	●
7 WIDE pH TOLERANCE	●
8 PERMANENTLY WET	●
9 HEAVY, MOISTURE RETAINING	●
10 MEDIUM LOAM	●
11 LIGHT, FREE-DRAINING	●
12 INFERTILE SOIL TOLERANT	●
ASPECT	
13 HARDINESS 1, 2, 3	3
14 EXPOSURE: COASTAL (SALT TOLERANT)	●
15 EXPOSURE: INLAND	●
16 SHADE TOLERANT	ST
PERFORMANCE	
17 GROWTH RATE	
SLOW	S
MEDIUM	M
FAST	F
18 EFFECTIVE LIFE	
SHORT	S
MEDIUM	M
LONG	L
19 PEST/DISEASES	
HIGHLY SUSCEPTIBLE	●
MODERATELY SUSCEPTIBLE	○
20 TOUGHNESS: SURVIVOR/PIONEER	●
21 SHELTER/SCREEN	●
22 HEDGE	●
23 GROUND COVER	●
24 BROAD SPREAD	●
25 NARROW SPREAD	●
26 SPIKE	●
27 COLUMNAR	●
28 PYRAMIDAL	●
29 OVERALL FOLIAGE COLOUR	
YELLOW/GOLD	YL/GD
PURPLE/BLUE	PR/BL
SILVER/GREY	SV
VARIEGATED	VR
30 WINTER COLOUR CHANGE	●

Fig. 7.2 *(Contd)*

TREES AND SHRUBS FOR LANDSCAPE PLANTING

SECTION 2 SHRUBS

GENERA	species 'Variety' or 'Cultivar'	"English name"	1 LARGE	2 MEDIUM	3 SMALL/PROSTRATE	4 EVERGREEN / SEMI EVERGREEN	5 DEMANDS ACID SOIL	6 WIDE PH TOLERANCE	7 PERMANENTLY WET	8 HEAVY, MOISTURE RETAINING	9 MEDIUM LOAM	10 LIGHT, FREE DRAINING	11 INFERTILE SOIL TOLERANT	12 HARDINESS 1, 2, 3	13 EXPOSURE: COASTAL (SALT TOLERANT)	14 EXPOSURE: INLAND	15 MAXIMUM SUN: S. WALL	16 FULL SUN PREFERRED	17 SHADE TOLERANT / HEAVY OVERHEAD SHADE
			HABIT				SOILS							ASPECT					
LAVANDULA	angustifolia 'Grappenhall' (L. spica in part)				●	E		●			●	●	●	3	●	○		●	
"Lavender"	angustifolia 'Hidcote'				●	E		●			●	●	●	3	●	○		●	
	angustifolia 'Twickel Purple'				●	E		●			●	●	●	3	●	○		●	
LEUCOTHOE	fontanesiana (catesbaei)				●	E	●			●	●	○		3					●
LEYCESTERIA	formosa		●					●		●	●	●		3	○	○			○
LIGUSTRUM	lucidum		●			E		●			●	●		2					○
	ovalifolium	"Oval-leaf Privet"	●			E		●		●	●	●	●	3	●	●			●
	ovalifolium 'Aureum'	"Golden Privet"		●		E		●		●	●	●	●	3	●	●			●
	vulgare	"Common Privet"		●		S/E		●			●	●	●	3	●	●			●
LONICERA	nitida 'Baggesen's Gold'			●		E		●		○	●	●	○	3		●			○
	nitida 'Ernest Wilson' (nitida HORT)			●		E		●		○	●	●	○	3		●			○
	pileata				●P	E		●		○	●	●		3	○	○			○
	x purpusii 'Winter Beauty'			●				●		○	●	●		3					○
	syringantha			●				●		○	●	●		3		●			○
	tatarica 'Hack's Red'			●				●		○	●	●		3		●			○
MAGNOLIA	grandiflora 'Exmouth'		●			E		○		○	●	○		2	○		○	●	
	liliiflora 'Nigra'			●			○			○	●	○		3					○
	loebneri 'Leonard Messel'		●					○		○	●	○		3					○
	loebneri 'Merrill'		●					○		○	●	○		3					○
	x soulangiana		●					○		○	●	○		3					○
	x soulangiana 'Alba Superba'		●					○		○	●	○		3					○
	x soulangiana 'Lennei'		●					○		○	●	○		3					○
	x soulangiana 'Rustica Rubra'		●					○		○	●	○		3					○
	stellata			●			○			○	●	○		3					○
MAHONIA	aquifolium	"Oregon Grape"			●	E		●		○	●	●		3		●			●
	aquifolium 'Apollo'				●	E		●		○	●	●		3		●			●
	'Charity'		●			E		●		○	●	●		3					●
	japonica			●		E		●		○	●	●		3					●
	'Lionel Fortescue'		●			E		●		○	●	●		3					●
	pinnata			●		E		●		○	●	●		3					●
	'Undulata'			●		E		●		○	●	●		3		○			●
MYRTUS	communis	"Myrtle"	●			E		●			●	●		1	○		●	●	
NEILLIA	thibetica (longiracemosa)			●				●		●	●	●		3					○
OLEARIA	x haastii			●		E		●			●	●		3	●	●		●	○
	macrodonta 'Major'			●		E		●			●	●		2	●			●	○
OSMANTHUS	delavayi			●		E		●			●	●		3					○
X OSMAREA	'Burkwoodii'			●		E		●			●	●		3					○
PACHYSANDRA	terminalis				●P	E		○		●	●	○		3		●			●

Fig. 7.2 (*Contd*)

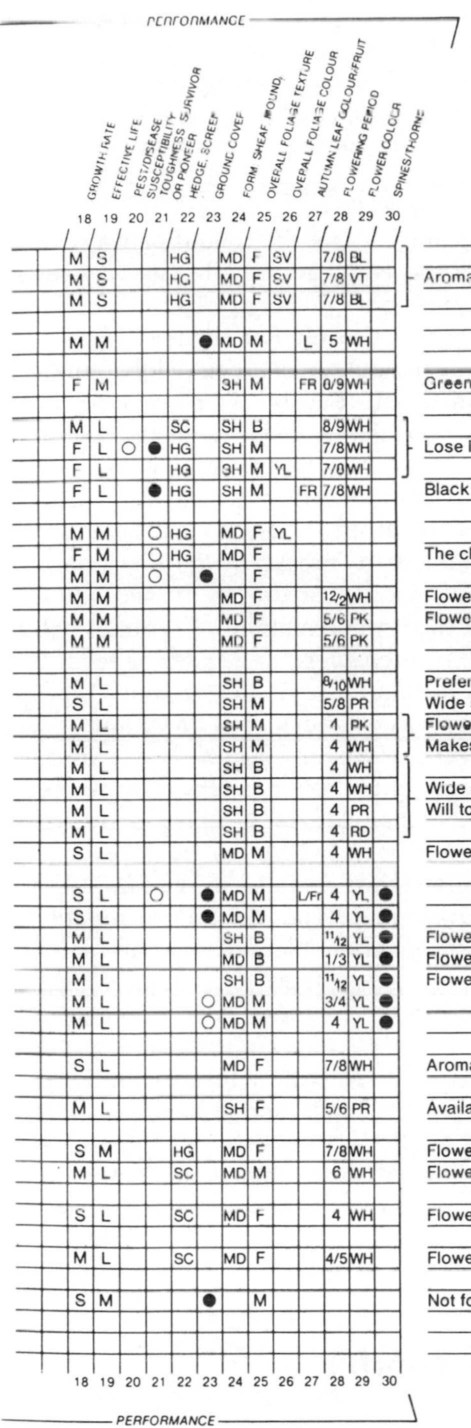

PERFORMANCE

18 GROWTH RATE	19 EFFECTIVE LIFE	20 PEST/DISEASE SUSCEPTIBILITY	21 TOUGHNESS: SURVIVOR OR PIONEER	22 HEDGE, SCREEN	23 GROUND COVER	24 FORM: SHEAF, MOUND	25 OVERALL FOLIAGE TEXTURE	26 OVERALL FOLIAGE COLOUR	27 AUTUMN LEAF COLOUR/FRUIT	28 FLOWERING PERIOD	29 FLOWER COLOUR	30 SPINES/THORNS	OTHER FACTORS
M	S			HG		MD	F	SV		7/8	BL		
M	S			HG		MD	F	SV		7/8	VT		Aromatic foliage/flowers
M	S			HG		MD	F	SV		7/8	BL		
M	M				●	MD	M		L	5	WH		
F	M					SH	M		FR	8/9	WH		Green stems; flowers red bracts.
M	L			SC		SH	B			8/9	WH		
F	L	○	●	HG		SH	M			7/8	WH		Lose leaves in cold winters.
F	L			HG		SH	M	YL		7/8	WH		
F	L		●	HG		SH	M		FR	7/8	WH		Black berries
M	M		○	HG		MD	F	YL					
F	M		○	HG		MD	F						The clone now normally grown.
M	M		○		●		F						
M	M					MD	F			12/2	WH		Flowers scented – winter
M	M					MD	F			5/6	PK		Flowers scented
M	M					MD	F			5/6	PK		
M	L					SH	B			8/10	WH		Prefers warm wall; scented flowers
S	L					SH	M			5/8	PR		Wide spreading.
M	L					SH	M			4	PK		Flowers scented. Stellate.
M	L					SH	M			4	WH		Makes a small tree.
M	L					SH	B			4	WH		
M	L					SH	B			4	WH		Wide spreading.
M	L					SH	B			4	PR		Will tolerate some alkalinity.
M	L					SH	B			4	RD		
S	L					MD	M			4	WH		Flowers scented. Stellate.
S	L		○		●	MD	M		L/Fr	4	YL	●	
S	L				●	MD	M			4	YL	●	
M	L					SH	B			11/12	YL	●	Flowers scented
M	L					MD	B			1/3	YL	●	Flowers scented
M	L					SH	B			11/12	YL	●	Flowers scented
M	L				○	MD	M			3/4	YL	●	
M	L				○	MD	M			4	YL	●	
S	L					MD	F			7/8	WH		Aromatic foliage, flowers scented.
M	L					SH	F			5/6	PR		Availability factor.
S	M			HG		MD	F			7/8	WH		Flowers scented
M	L			SC		MD	M			6	WH		Flowers scented. Seaside shelter.
S	L			SC		MD	F			4	WH		Flowers scented, clips well.
M	L			SC		MD	F			4/5	WH		Flowers scented
S	M				●		M						Not for shallow, chalky soils.

18 19 20 21 22 23 24 25 26 27 28 29 30 — OTHER FACTORS

PERFORMANCE

Fig. 7.2 (*Contd*)

KEY TO SYMBOLS AND ABBREVIATIONS FOR SHRUB SECTION

● = PLANT FALLS CLEARLY INTO CATEGORY
○ = PLANT MAY BE SUITABLE FOR CATEGORY

HABIT
1 LARGE ●
2 MEDIUM ●
3 SMALL/PROSTRATE ● P
4 EVERGREEN E
SEMI EVERGREEN SE

SOILS
5 DEMANDS ACID SOIL ●
6 WIDE pH TOLERANCE ●
7 PERMANENTLY WET ●
8 HEAVY, MOISTURE RETAINING ●
9 MEDIUM LOAM ●
10 LIGHT, FREE-DRAINING ●
11 INFERTILE SOIL TOLERANT ●

ASPECT
12 HARDINESS 1, 2, 3 3
13 EXPOSURE: COASTAL (SALT TOLERANT) ●
14 EXPOSURE: INLAND ●
15 MAXIMUM SUN, S.WALL ●
16 FULL SUN PREFERRED ●
17 SHADE TOLERANT ○
HEAVY OVERHEAD SHADE ●

PERFORMANCE
18 GROWTH RATE:
SLOW S
MEDIUM M
FAST F
19 EFFECTIVE LIFE:
SHORT S
MEDIUM M
LONG L
20 PEST/DISEASES:
HIGHLY SUSCEPTIBLE ●
MODERATELY SUSCEPTIBLE ○
21 TOUGHNESS: SURVIVOR/ PIONEER ●
22 HEDGE HG
SCREEN SC
23 GROUND COVER ●
24 FORM:
SHEAF/VASE SHAPED SH
MOUND/DOME MD
25 OVERALL FOLIAGE TEXTURE:
FINE TEXTURE F
MEDIUM TEXTURE M
BOLD TEXTURE B
26 OVERALL FOLIAGE COLOUR:
YELLOW YL
PURPLE PR
SILVER/GREY SV
VARIEGATED VR
27 AUTUMN LEAF COLOUR L
FRUIT FR
28 FLOWERING PERIOD 7-10
29 FLOWER COLOUR:
WHITE WH
PINK PK
RED RD
VIOLET VT
BLUE BL
YELLOW/CREAM YL
ORANGE OR
30 SPINES OR THORNS ●

selecting from the 12 000 plants that can be grown in this country, over 750 herbaceous and 100 trees and shrubs which should usually be readily available.

In the trees and shrubs list, those that may be in short supply due to difficulties of propagation, those raised from seed too quickly which reach saleable size in the nursery and then tend to deteriorate, those that are too tender for exposed positions in all but the mildest localities, and those that will not tolerate alkaline or chalky conditions, are all indicated separately.

In choosing from the herbaceous plant list, specifiers must always remember that should the microclimate of the proposed location be in any way susceptible to frost the following five species will always be at risk: *Choisya ternata, Eucalyptus gunii, Hebe* spp., *Senecio greyii*, and *Garrya eliptica*.

Specification of those plants available in containers are laid down in the British Container Growers Group List, which sets out the standards to which the majority of container-grown plants are available, giving both the minimum and recommended capacities of the containers and/or the minimum heights or the diameters of the plant (Figure 7.3). It also describes the physical shape of the plants when offered for sale: branched or bushy, single leader, side shoots or, if multi-stemmed, with details of the number of breaks in the lower third of the plant: whether terminals and/or laterals should have been pruned and if caning is to be provided. Details are given of the normal method of raising: by budding, stem cutting, division, graft, layering, root cuttings, by runner or seed. It is thus possible for the specifier to be confident that not only the size but also the shape and method of propagation of container grown stock are known and ensured for the more common cultivars. In general it should always be borne in mind that most plants grown in containers are unlikely to be safely transplanted if the containers in which they are sold are less than 3 litres; 4–5 litres are usually more advantageous, and in some circumstances even 7–10 litres are necessary.

The success of any landscape project stands or falls on the supply to site of good quality healthy plants by the nurseryman. The well-known reputable firms appreciate this and take great care in ensuring that their plants are properly protected in transit. There are unfortunately others who buy in plants from wherever the cheapest can be found irrespective of their condition and take little interest or care in their delivery to site.

Severe competition today requires landscape contractors to seek keener and keener markets to reduce their costs. In the landscape industry more than any other you get what you pay for. Good quality plants cost more than inferior ones and it is essential that landscape architects ensure that their plants come from reputable sources at realistic prices if they are to get the quality they are looking for.

7.2 JLCPS/CPSE Standard Form of Tender

The CPSE (previously JLCPS) Standard Form of Tender for the Supply and Delivery of Plants (Appendix E) lays down a standard format for the specification

THE BRITISH CONTAINER GROWERS GROUP

Specification of Standards for the Production of Hardy Container Grown Plants

---

EDWARD BACK	FARGRO LTD, CANTERBURY ROAD, WORTHING
PETER BULLOCK	J. & W. BLACKBURN LTD., SHELLEY, HUDDERSFIELD
DAVID CLARK	NOTCUTTS NURSERIES LTD, WOODBRIDGE, SUFFOLK
MIKE CLIFT	WATERERS NURSERIES, BAGSHOT, SURREY
LIN DEMPSTER	BOULTON BROS. MODDERSHALL LTD., STONE, STAFFS.
MIKE DUNNETT	BLAKEDOWN NURSERIES LTD., KIDDERMINSTER, WORCS.
BRIAN HUMPHREY	HILLIERS NURSERIES (WINCHESTER) LTD., HAMPSHIRE
BARRY LOCKWOOD (Chairman)	ROSES AND SHRUBS LTD., WOLVERHAMPTON, STAFFS.
ROBIN TACCHI	TACCHI'S NURSERIES, HUNTINGDON
	MIDLAND NURSERIES LTD., SOLIHULL, WARWICKSHIRE
	BERNHARDS RUGBY NURSERIES LTD., RUGBY, WARWICKSHIRE.

Fig. 7.3 Specification of standards for container-grown plants, issued by the British Container Growers Group.

INTRODUCTION

This document is intended to provide a definition of standards to which a range of Container Grown plants should be produced.

The majority of plants produced in quantity are represented, if not specifically, then by a comparable plant.

COLUMN No. 1 **PLANT GENERA**–represented with additional vigor categories for larger Genera.

COLUMN No. 2 **'MINIMUM STANDARD'** – below which stock should not be offered. The figures are shown thus – e.g. 2/30. The first figure 2 indicates the volume capacity of the container in litres, followed by the minimum height of the plant in centimetres, measured from the surface of the compost. 'D' refers to the diameter measurement where applicable.

COLUMN No. 3 The **'RECOMMENDED STANDARD'** is stated also in Container Volume and Height. It is proposed that this column should be the standard to which growers should progress and would, in time, replace the 'Minimum Standard Column.'

COLUMN No. 4 **'PLANT HABIT'** – describes the physical appearance of the plant with the following definitions:

Branched – A branched plant with few shoots arising from the branches.

Bushy – A branched plant with several shoots arising from the branches.

Leader and Laterals – A single dominant shoot with significant side shoots.

Single Leader – A single dominant shoot with few, if any, side shoots.

Several Shoots – A number of shoots arising from ground level.

COLUMN No. 5 **NUMBER OF BREAKS IN LOWER THIRD OF THE PLANT**

COLUMN No. 6 **PRUNING CODE** – refers to the type of pruning required to achieve the required plant habit.

1. Prune where required.
2. Terminal pruned.
3. Laterals pruned.
4. Terminal and Laterals pruned.

COLUMN No. 7 **CANING**

COLUMN No. 8 **NORMAL METHOD OF RAISING**

B – BUDDING
C – STEM CUTTINGS
D – DIVISION
G – GRAFTS
L – LAYERS
R/C – ROOT CUTTINGS
R – RUNNER
S – SEED

COLUMN No. 9 **EXAMPLE** – provides a representative specific plant of the Genera, to which the specification applies.

C.V.'s – Cultivars.

Fig. 7.3 (*Contd*)

BRITISH CONTAINER GROWERS GROUP

Specification of Standards for the Production of Hardy Container Grown Plants

PLANT	Minimum Standards Vol. in Litres Hgt. in Cms.	Recommended Standards Vol. in Litres Hgt. in Cms.	Plant Habit	No. of breaks in lower ⅓	Pruning Code	Caning	Normal Method of Raising	EXAMPLE
1	2	3	4	5	6	7	8	9
SHRUBS								
Abelia	1/20	2/25	Bushy	3	2	–	C.	A. grandiflora
Acacia	2/30	4/40	Single leader	–	1	–	S.	A. dealbata
Acer Medium	3/30	4/40	Branched	2	1	–	C.S.L.G.	A. palmatum atropurpureum
,, Small	2/20	3/25	Branched	2	1	–	C.S.L.G.	A. jap. 'Aureum' A. palmatum Dissectum CVS.
Aesculus Small	4/20	6/30	Branched	2	1	–	S.G.	A. parvifolia
Amelanchier	3/30	4/40	Branched	3	1	–	S.L.G.	A. lamarkii (canadensis) A. canadensis
Aralia	3/30	4/40	Single leader	–	1	–	S.R/C.G.	A. chinensis
Arbutus	2/20	3/30	Branched	2	1	–	S.	A. unedo
Aronia	2/30	3/40	Branched	3	1	–	C.S.	A. melanocarpa
Arundinaria	3/40	4/60	Several shoots	3	1	–	D.	A. mureliae
,, Dwarf	1/15	2/25	Several shoots	3	1	–	D.	A. humilis
Artemisia	1/15	2/20	Bushy	3	1	–	C.	A. abrotanum
Atriplex	1/20	2/30	Bushy	3	2	–	C.S.	A. halimus
Aucuba	2/20	3/30	Bushy	3	2	–	C.	A. japonica
Azalea *see Rhododendron section*								
Azara	1/20	2/30	Leader and Laterals	3	1	1	C.	A. microphylla
Berberis Evergreen Vigorous	3/30	4/40	Bushy	3	4	–	C.S.	B. stenophylla
,, ,, Medium	2/20	3/30	Bushy	3	4	–	C.S.	B. darwinii
,, ,, Dwarf	2/15D	3/20D	Bushy	3	1	–	C.S.	B. candidula
Berberis Deciduous Vigorous	3/30	4/40	Branched	3	4	–	S.C.	B. Ottowensis C.V.S.
,, ,, Medium	2/20	3/30	Branched	3	4	–	S.C.	B. thunbergii CVS.
,, ,, Dwarf	1/10	2/15	Bushy	3	4	–	C.	B. thunbergii atro. Nana
Buddleia	2/30	3/40	Branched	3	4	–	C.	B. Davidii & CVS.
Buxus Medium	2/20	3/25	Bushy	3	1	–	C.	B. sempervirens CVS.
,, Dwarf	1/10	2/15	Bushy	3	1	–	C.	B.s. suffruticosa
Callicarpa	2/30	3/40	Branched	3	2	–	C.	C. giraldiana Profusion
Calluna *see Erica*								
Camellia	1.5/30	3/40	Leader with Laterals	–	1	-	C.G.	C. japonica CVS.
Caryopteris	1/15	2/20	Branched	3	2	–	C.	C. clandonensis
Ceanothus (*Also climbing and wall plants*)	1.5/45	3/60	Leader with Laterals	–	1	C.	C.	C. burkwoodii
Ceratostigma	1/15	2/20	Bushy	3	1	–	C.	C. willmottianum
Cercis	3/30	4/40	Branched	3	1	–	C.S.	C. siliquastrum
Chaenomeles	2/30	3/40	Branched	2	1	Cane C.V.'s	C.	C. japonica CVS.
Chimonanthus	2/30	3/40	Branched	3	2	–	S.C.G.	C. fragrans
Choisya	2/20	3/30	Bushy	2	2	–	C.	C. ternata
Cistus	1/20	2/30	Bushy	3	1	–	C.	C. Silver Pink
Clerodendron	3/30	4/40	Branched	2	1	–	R/C.S.	C. trichotumum
Clethra	2/20	3/30	Branched	3	1	–	C.S.	C. alnifolia
Convolvulus	0.5/10D	1/15D	Bushy	3	1	–	C.	C. cneorum
Colutea	3/30	4/40	Branched	3	2	–	S.C.	C. arborescens
Cornus Vigorous	3/30	4/40	Branched	3	2	–	C.	C. alba & CVS.
,, Medium	2/30	3/40	Branched	3	2	–	L.C.S.	C. florida 'Rubra'
,, Dwarf	0.5/10D	1/15D	Several shoots	3	1		D.	C. canadensis

Fig. 7.3 (*Contd*)

of plants and for tenders for the supply of all trees, shrubs and herbaceous plants, whether container grown or not. It contains standard details, conditions, specification, schedules and a form of tender for the supply of all types of plants to statutory authorities and local authorities in the public sector, and landscape architects and managers acting on behalf of all types of client in the private sector alike.

The form is primarily intended for all types of enquiry for the supply of plants including those for contracts under the JCLI Form. When the form is used to obtain prices for a range of plants only certain species of which are to be selected, it is essential that this is clearly stated in the enquiry with the minimum quantity of any one species anticipated, if the prices quoted are not to be valueless. Several points should be noted, as detailed below.

It is appreciated that the purchaser needs to retain the right to accept orders in whole or in part, but in the case of the latter (partial orders) it is accepted that the supplier has the right to re-negotiate rates where the quantity or variety ordered differs significantly from the quantity specified, or the value of the order placed differs significantly from the total tender.

Specifiers are reminded that some bushy plants are only available in small sizes and conifers are rarely available as standards.

The origin of plants other than those grown in the British Isles should be identified by the supplier against the names of relevant plants in the schedules.

Root systems are not specified in detail other than that they must be adequate in relation to the size of the plant, conducive to successful transplanting, and that trees are to be bottom-worked unless otherwise stated.

Any container-grown plants required must be so indicated with the minimum size of container stated, the roots to properly permeate the whole of the container without being pot-bound or rooted through.

Quantities (contrary to BS 1192) must be priced, monied out, and totalled, with large quantities rounded off (transplants usually being in units of 25). Trees, shrubs, roses and herbaceous plants should be listed separately in alphabetical order to assist the estimator, leaving space below each plant for substitutes to be entered when these are allowed by the specifier. When plants are no longer available at the time of delivery, substitutes may be made only by mutual agreement. It is helpful if two copies of each schedule are sent with the enquiry so that the estimator can return the original and retain a copy for his file.

No plants are to be supplied until the acceptance of the tender has been confirmed in writing.

A minimum of two clear working weeks, with three weeks wherever possible, is the recommended time for tendering, with the closing date for receipt of tenders clearly shown on the reply envelope where these are provided.

Although not stated on the form it is suggested that tenders should remain open for acceptance for 28 days during which time prices will remain firm and stocks be guaranteed but beyond which prices may be subject to increase and stock of plants in the quantities specified may no longer be available.

The use of this form of tender ensures comparability of tenders so that like can be compared with like and all the plants supplied comply with the minimum standards laid down (Figure 7.4, for example). Although not always practicable it is strongly recommended that wherever possible the plants of the lowest tenderer are inspected in the nursery by prior appointment before the order is placed.

Delivery dates (the dates by which, or on which, deliveries are required) must be stated in the enquiry and are to be adhered to, save only delays on account of exceptionally adverse weather conditions or other reasons beyond the supplier's control. The supplier is required to advise the purchaser in advance of the week during which deliveries are intended, and, where sites are identified on the Form of Tender by the specified as not being continually manned, the supplier is required to give 48 hours' notice of the date and anticipated time of arrival on site so that adequate labour for unloading and checking can be provided by the purchaser. Deliveries are to be made during normal working hours and, except where stated otherwise, to one location.

Complaints regarding quantity, quality and condition should always be made within seven days of delivery, which may be signed for as 'unexamined', the purchaser having the right to return at the suppliers' expense any plants the quality of which is below that which was specified.

Payment should always be made by the end of the month during which deliveries are made notwithstanding that separate invoices may be rendered in the event of deliveries being made in separate consignments. Prices are usually inclusive of delivery and are strictly net with no discount for prompt payment, exclusive of value added tax which will be charged at the rates ruling at the time of delivery.

Postponement of deliveries until the following planting season after 31 March by the purchaser entitles the supplier to the reimbursement of the direct loss and expenses to which he has been put by the postponement.

Should it not be possible to resolve any dispute regarding an order based on the JLPSE/CPSE conditions either party has the right to refer the matter to an arbitrator appointed by the President of the Institute of Arbitrators on the request of any of the organizations named on the front of the form.

The Form of Tender requires the specifier to complete at the head of the form, in addition to his own name and address, the name and address of the project, the address to which delivery is to be made, and to state the date and time when deliveries are to be made, especially for those sites which will not be continuously manned, together with the latest date and time for the receipt of the tender. The form also provides for the option of deleting the obligation to deliver the plants to site for those specifiers who prefer to collect them themselves.

Space then follows on the form for the supplier to summarize his tender, adding the cost of any non-standard special conditions imposed by the specifier, his delivery charges, and the total amount of his tender (exclusive of the value added tax to be added at the rates ruling at the date of delivery) in figures and in words.

The supplier also undertakes (subject to their being unsold on receipt of order) to supply the plants at the price quoted during the current bare-root planting season subject to the order being placed within four weeks of the date for submission of tenders.

Dimensions of Trees

Designation	Circumference of stem measured 1.0m from ground level	Minimum Height from ground level	Clear stem Ht. (m) from ground level to lowest branch
BS 3936 Part 4 Specification for Forest Trees.			
Younger nursery stock			
Seedlings (1+0)			
(2+0)		max. 0.90 m	—
(etc.)			
Undercut (1+1)			
Seedling/(1+2)	not specified	max. 0.90 m	not specified
Transplant (etc.)			
Whip		0.90–1.20 m	not specified
		1.20–1.50 m	
		1.50–1.80 m	
		1.80–2.10 m	
		2.10–2.50 m	
BS 3936 Part 1.			
BS 3936 Part 1 Standard Nursery Stock . Standard/feathered/weeping			
short	not specified	not specified	1.00–1.20 m
half	not specified	1.80–2.10 m	1.20–1.50 m
	4–6 cm	2.10–2.50 m	1.50–1.80 m
Light	6–8 cm	2.50–2.75 m	1.50–1.80 m
Standard	8–10 cm	2.75–3.00 m	1.80 minimum
Tall	8–10 cm	3.00–3.60 m	1.80 minimum
Selected	10–12 cm	3.00–3.60 m	1.80 minimum
Note: Minimum heights do not apply to weeping trees nor clear stem height to feathered trees.			
BS 5236 Advanced Nursery Stock			
Heavy	12–14 cm	3.60–4.25 m	1.80 minimum
Extra heavy	14–16 cm	4.25–6.00 m	1.80 minimum
	16–18 cm		1.80 minimum
	18–20 cm		1.80 minimum
Note: Minimum height of conifers 2.0–3.5 m			
(BS 4043) Semi Mature	20–75 cm	6.00–15.00	not specified

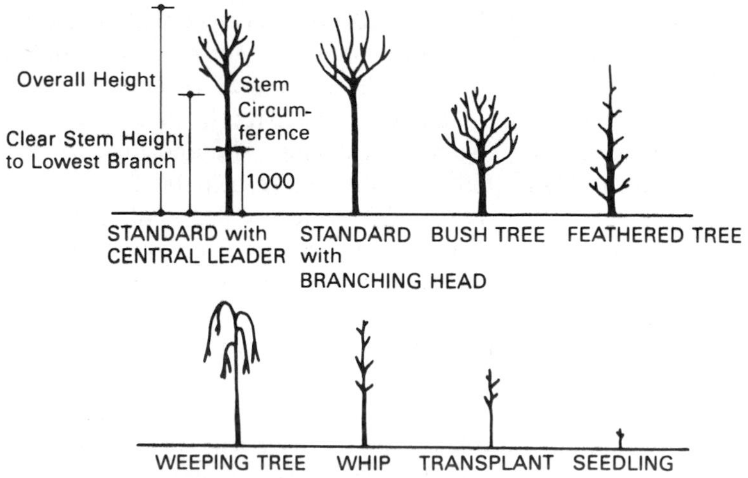

Fig. 7.4 Dimensions of trees – some standards quoted.

The specifier does not bind himself to accept the lowest (or any) tender, and the supplier retains the right to withdraw his offer if the plants are to be supplied through a contractor whose identity is not known to him at the time of his tender, within 14 days of the identity of the contractor becoming known.

7.3 JCT Form of Tender for nominated suppliers

Just as landscape subcontractors on building contracts have to comply with obligatory subcontract provisions and procedures to ensure compliance with the conditions of the main contract, landscape suppliers on building contracts have similar special conditions.

In this instance, however, the requirements are in some respects far less onerous: not only is there no restriction on naming a single supplier in the tender and contract documents (provided that the work is described in sufficient detail for it to be priced by the main contractor) but neither is the form of tender and employer subcontractor agreement obligatory.

Should, however, the work not be fully described and a prime cost or provisional sum be provided instead, then the supplier is considered to have been nominated and the detailed provisions then come into force. Since, under these circumstances, the main contractor can automatically obtain an extension to the contract period and the employer lose the right to liquidated and ascertained damages if the main contractor is delayed by the nominated supplier, it is essential that a contractual situation is created between the employer and nominated supplier by the use of a standard JCT Employer/Supplier Agreement similar to that available for subcontractors.

Form of Tender and Agreement (TNS/1)

Although this does not contain the detailed provisions (with regard to the specification, packaging and delivery of plants to site) set out in the CPSE Form, the JCT Form of Tender imposes different if equally onerous obligations. The Form of Tender (Appendix F) reprints in full the voluminous provisions of JCT clause 36.4 which requires the architect only to nominate a supplier who will agree to enter into a contract to supply the goods required on the same terms as the main contract (unless the main contractor agrees otherwise). In practice this does not create any problems, with two exceptions.

1. Clause 36.4.7 requires that 'The ownership of materials or goods shall pass to the main contractor upon delivery . . . whether or not payment has been made in full'. This in effect precludes the architect from nominating, without the consent of the contractor, any supplier who operates a retention-of-title clause of any kind.

2. In addition, the Employer/Supplier Agreement incorporated into the Form of Tender obliges the supplier under clause 1.2 to '. . . commence and complete delivery in accordance with the arrangement of our contract of sale . . . so that the contractor shall not be entitled to an extension of time'. In default of which, should it arise, the employer is entitled to recover from the nominated supplier the full liquidated damages to which he is entitled under the main contract – which may well be, on a multi-million-pound contract, a sum vastly in excess of the value of the goods to be supplied. The only proviso is that if delivery by the supplier has been delayed by (a) *force majeure*, (b) civil commotion, strike or lock out, (c) architect's instructions, or (d) failure of the architect to supply necessary instructions in time, situations unlikely to be of much relevance in connection with the supply of plants.

The JCT Form TNS/1 also deals with the obligation to exercise 'reasonable skill and care in the design of any of the goods to be supplied'. This is of little relevance to plants, particularly those whose ultimate destination and location is unknown to the supplier at the time of tender.

Under any form of contract or tender, the supply of plants is also subject to the full provisions of the Sale of Goods Act 1979 in which Section 14 requires that:

1. the goods will comply with the description detailed (specified);
2. the goods are of merchantable quality except
   (a) where attention is drawn to the fault at the time of sale, or
   (b) the buyer examines the goods before the contract is made*;
3. The buyer expressly or by implication states that the goods specified are reasonably fit for the purpose intended.

* This means that under the Act inspection in the nursery relieves the supplier of his obligations to provide that which was specified.

8

Plant handling

8.1 Introduction

Having obtained comparable tenders for the provision of plants of adequate minimum standards of quality, there remains the problem of ensuring their survival when lifted, packaged and transported to site by the nursery, and their subsequent handling and establishment on site where the efforts of the landscape contractor, subcontractor or purchaser are equally of crucial importance.

The survival and establishment of the plants specified should be the primary aim of every landscape architect, contractor, subcontractor and supplier. The efforts of the nurseryman to produce healthy and vigorous plants are worthless if the plants are put needlessly at risk during the period in which they are lifted and transported from the nursery to their intended new location, and while awaiting planting on site.

Meticulous care and attention to detail is essential during this period and involves the following.

1. The highest standards of plant handling within the nursery. This is almost entirely dependent on the standards set by the growers themselves since plant purchasers are rarely able to be present and inspect this work while it is being carried out.
2. A stringent specification for packaging while in transit, to be set out in tender documents and rigidly enforced by the purchaser; any plants not packaged in accordance with the minimum standards laid down to be rejected on delivery from the nursery.
3. A code of good practice for the handling of plants on site which, if followed, should ensure the successful establishment of healthy plants and which, if it

were found by the supplier not to have been adhered to, would relieve him of further responsibility should the plants subsequently die.

Poor plant handling between the nursery bed and the final planting position, however healthy and carefully looked after the plant may be, must inevitably contribute to the majority of plant losses. The JLCPS/CPSE Code of Practice for Plant Handling (Appendix E) was first published in 1980, but early editions failed to make compliance with its requirements obligatory and, concentrating on minimizing losses from desiccation, omitted detailed provisions for the protection of roots, and failed to differentiate between those items such as packaging and bundling which could be enforced by rejection on delivery and those items which were dependent on the concern of the nurseryman for his plants before they left his premises.

8.2 Handling within the nursery

Once out of the ground, plants are highly susceptible to damage especially of their roots, which must therefore be exposed to damage for as short a time as possible. This applies to all plants – trees, shrubs or herbaceous – and every effort must be made to protect the roots and prevent them drying out, heating up, freezing, becoming water logged or broken. These risks apply equally to container-grown and bare-root plants lifted and despatched during the dormant season, although the risks of damage to container-grown plants, while they still exist, are clearly less.

Time of lifting

Bare-root and root-balled plants must only be lifted when the ground is moist and the plant is dormant, usually between October and March for deciduous plants, and September to May for evergreens, dependent on local and seasonal weather conditions. Particular attention must always be paid to the protection of the roots on lifting when there is a strong drying wind or sun; it should never take place when there is a severe ground frost. If plants are to be delivered with the roots balled (which includes all evergreen plants over three years old) these must be adequately supported to prevent collapse.

Container grown plants can be moved at any time of the year provided that the root system is well established and the plants are fully hardened off.

Bundling

This must only take place when the foliage of evergreen plants is reasonably surface dry if heating up, with consequent damage to the plant, is not to occur.

Bundles of bare root plants should be of a manageable size. Root-balled and container-grown plants must never be bundled.

Labelling

If plants are not labelled prior to lifting care must be taken to ensure that all species are kept separate and identifiable until such time as the labels can be provided.

Similar species such as *Alnus* whips (*A. cordata, A. glutinosa, A. incana*) or varieties of *Potentilla* must always be kept separately identifiable, in each and every bundle.

Protection

Care must be taken to ensure that bare roots are protected from physical damage and desiccation at all times, and that, when in an open vehicle, they are securely sheeted over. All bare roots must be covered within two hours of lifting if irreparable damage to the plant is to be avoided. Ball-rooted and container-grown plants may be transported within the nursery without additional protection unless they are likely to be exposed to extreme or sudden changes of temperature for more than a short period of time.

Temporary storage within the nursery

1. **Bare-root plants**
   (a) **Externally** Plants may be kept outside in a moist, cool, sheltered and shaded location for several weeks in the nursery, provided that the roots are surrounded with a freely draining moisture-retentive material, moistened regularly as and when necessary. It is essential that good contact is ensured between the material and the roots, and that protection against damage from small rodents is provided.
   (b) **Internally** Exposure for a relatively short period even in an unheated packing shed (let alone a garden centre) can result in serious drying out and can result in the loss of the plant. The roots of plants exposed to such risks for more than a few hours must be frequently watered. If they are likely to be exposed for longer periods, they must be protected and maintained in a similar way to that laid down for temporary storage externally.
   (c) **In cold stores** Plants must be wrapped entirely in polythene unless they are packed or humidified in which case no further protection is necessary.

2. **Root-balled plants**
   Where roots are protected in this way, they can be kept for short periods provided permeable wrappings are kept moist by watering, and polythene wrappings are kept from direct sunlight. For longer periods, permeable wrappings must be placed on a concrete or well-drained surface and covered with the same freely draining moisture-retentative material, moistened as necessary and protected from small rodents in the same way as bare-root plants.
3. **Container-grown plants**
   These have only to be kept upright, watered as necessary and protected from strong winds and extremes of temperature.
4. **Transport**
   Moving plants around the nursery requires the same precautions and protection from damage as are laid down for plants in transit to the purchaser.

8.3 Plants in transit from the nursery to site

While the preceding recommendations for the lifting, handling, packaging and transportation of plants within the nursery are dependent on the standards set and the self-policing of the nurseryman, which can rarely be supervised by the purchaser, the following requirements can be laid down by the purchaser in his tender documents, and any shortfall from the standards of packaging specified can be immediately seen at the time of delivery to site. Any plants whose subsequent survival has been needlessly placed at risk as a result can be instantly identified and such plants be rejected and returned to the supplier at his own cost without argument.

Lifting

Large plants should have their original orientation clearly marked on the stem.

Bundling

Plants must be in bundles of the same species and size, all shoots must face in the same direction so that roots and shoots are not in contact, and must be of equal numbers. Plants must be loosely tied in bundles of a size manageable by one person (unless written agreement has been obtained in advance that mechanical handling equipment will be available at the place and time of delivery). Bundles are to be securely tied with string, twine, plastic strip or other supple material which will not, by its nature or tension, cause damage to the plants.

Labelling

Each individual plant, bundle, bag or lot of one species of plant shall be labelled with a securely attached pruning label, clearly indicating the plant name, grade and quantity in the bundle or bag with the total quantity clearly displayed together with the supplier's name. In a large consignment of plants of the same species, a secondary label, easily related to the primary label may also be necessary.

Protection in transit

1. **Bare-root plants**
   Plants up to 45 cm tall shall be entirely but loosely enclosed in plastic film bags (250 grams, 65 micron) securely tied at the top. Thorny or very bushy plants need only have their roots enclosed, provided the plastic bag is securely tied around the stem.
   Plants between 45 cm and 60 cm tall shall have the roots totally enclosed in a plastic film bag of suitable size and thickness.
   Plants between 60 cm and 90 cm tall may be supplied bare root but the roots shall at all times be kept moist and protected from drying out. Layers of plants in transit must be sheeted over.
   Transplants from 90 cm to 120 cm tall should have their roots totally enclosed in plastic except when supplied for large-scale planting when exceptions may be acceptable to the purchaser, provided that this is clearly specified in the tender invitation.
   Whips and trees between 120 cm and 350 cm tall (whips and BS 3936 standard nursery stock) may be supplied bare root, provided that the roots are kept moist at all times and protected from drying out. Layers in transit should be sheeted over.
   Trees in excess of 350 cm tall (BS 5236 advanced nursery stock) may be supplied bare root provided that the roots are packed with pre-wetted moisture-retentive granular or fibrous material and enclosed in either a permeable material such as hessian which must be regularly watered or 500 gram/130 micron (minimum) plastic film.
2. **Root-balled plants (including BS 4043 semi-mature trees)**
   They must have the root balls kept moist, adequately supported with bands to prevent collapse, and protected against drying out by wrapping firmly with plastic film or moisture-retentive material firmly secured over the top of the root ball.
3. **Container-grown plants**
   These will not normally require additional packing but degradable pots shall be enclosed and firmly secured in 350 gram/65 micron polythene film.

Table 8.1 summarizes these requirements.

Table 8.1 Protection of plants and trees in transit

Description	Size		Max. number in bundle	Protection
	Girth (metres)	Height (metres)		
BS 3936: younger				totally enclosed
nursery stock		0.15–0.30		in plastic
Part 4 seedlings		0.30–0.45		
				roots enclosed
		0.45–0.60		in plastic
			unspecified	
		0.60–0.90		
				bare roots kept moist
		0.90–1.20		
		1.20–1.50		
		—		
Transplants BS 3936:		1.50–1.80		
standard nursery stock				
Part 1 whips				
Short	—	—	(25)	
Half	—	1.80–2.10	(20)	
—	4–6	2.10–2.50	15	
Lt/Stand	6–8	2.50–2.75	10	bare roots kept moist
Full	8–10	2.75–3.00	5	
Tall	8–10	3.00–3.50	5	
Selected	10–12	3.00–3.50	3	
BS 5236: advanced nursery stock				
Heavy	12–14	3.50–4.25		
Extra heavy	14–16	4.25–6.00		
	16–18		individual	root ball
	18–20			
Semi-mature	20.75	6.00–15.00		

8.4 Transport of plants

In open/closed lorries and containers

All plants are to be loaded, stacked and unloaded in such a way that breakage or crushing – by the weight of plants above, by the securing ropes or by personnel responsible for loading/unloading – will not occur. The plants shall be completely and firmly covered with opaque sheeting (laid with due regard to the direction of travel). Plants in polythene bags should be sheeted to be shaded from direct sunlight.

All plants to be loaded in such a way that breakage or crushing by the weight of plants above is avoided during loading, transit and unloading. Plants should not be kept in the dark for more than four days.

Transit by third parties

Where transit is entrusted to others (neither the purchaser nor supplier), not only should consignments be in manageable units, but they should be securely crated and packaged to withstand physical damage and desiccation during a possibly prolonged journey.

It is equally essential that the need for care in handling, separating multiple loads and the effects of weather and time of delivery, is impressed on non-specialist hauliers when they have to be used.

Reception

It is essential that appropriate staff and facilities are available for unloading on delivery without delay, to minimize damage on arrival. Where sites are only manned at certain specific times these (or the date and times of delivery) should be clearly stated in the tender documents. Alternatively the purchaser should arrange for collection himself.

Additional precautions

Variations to the above requirements may be appropriate in certain special circumstances which should be identified and clearly set out in the tender documents so that any additional costs will be allowed for by the tenderers.

8.5 Plant handling after delivery to site

Transplanting and re-establishment, notwithstanding the relatively benign British climate and soil conditions, involve a plant in considerable stress. If it is to be successful, it must be planned to satisfy the plant's basic biological requirements, irrespective of those of the client, the landscape architect and the contractor. Provided the nurseryman has complied with the foregoing requirements for lifting, packaging and transport, one more hurdle remains before the successful re-establishment of the plant can be ensured. This involves the caring, handling, temporary storage, transporting and planting of the nursery stock by the contractor or purchaser. Even if this is specified in detail it will be to no avail if blanket or standard specifications are issued irrespective of the particular requirements of the project (which is not to say, standard clauses should not be incorporated in the interests of an instant appreciation of the specific requirements and good practice). A relatively short lifting and planting season, the logistics of lifting and supplying large numbers of different species and the financial pressures to complete a project within a given contract period or financial year irrespective of unpredictable or unsuitable weather conditions, all add conflicting pressures on the ideal planting time. To ensure the certain survival of plants it is essential that:

1. planting is planned and material is ordered as far as possible in advance;
2. planting specifications are prepared in detail to suit the particular requirements of the site;
3. planting is carried out only by skilled operatives under adequate supervision;
4. aftercare is properly arranged and, where necessary, adequate finance is allocated for its provision.

8.6 Planting seasons

Irrespective of the time of delivery, bare-root plants will only survive if planted in their final location during their dormant period (usually between the beginning of October and the end of March). Early plantings before the end of the calendar year are always more successful than those carried out in the new year from January to the end of March. Planting in January or February is particularly dangerous in 'hostile sites'. Late planting of bare-root and root-balled stock is always vulnerable to desiccation from spring droughts and is to be avoided unless regular watering has been allowed for. Evergreens are best planted in early autumn or late spring. Container-grown plants, on the other hand, can be planted at any time of the year provided weather conditions are appropriate and they are regularly watered.

8.7 Temporary storage

Bare-root stock

If the ground or weather at the time of delivery are not suitable bare-root plants should always be stored temporarily on or off site in more appropriate conditions. For short periods of up to seven days plants can be kept at low temperatures of up to 5 °C in the packaging in which they have been delivered provided the roots are kept moist. For longer periods bare-root stock has to be heeled in. Bare roots must be kept moist and kept in contact with a moisture-retentive medium such as a 50/50 mixture of coarse and sand and peat or sawdust. Bundled plants must be cut open and spread out to ensure that such contact is achieved. Heeling-in sites must be well drained and sheltered, stout rails being provided to support trees of standard, advanced and semi-mature size. Hose watering points are always essential. Under such circumstances plants can remain heeled in for up to five months between November and March, but the shorter the period the better. It is essential to prevent the roots drying out when the plants are eventually lifted and transplanted to their final positions.

Root-balled trees

These must be equally well supported with the ball immersed in a deep layer of moist straw, hay, sand, peat, pulverized bark or other suitable material. Regular watering to prevent the root ball drying out is essential.

Container-grown stock

This must be stood on well-drained, level, weed-free ground and should be daily watered from April to October. Tall plants should be properly supported to prevent their being blown over.

8.8 Transporting to final location

The bare roots of plants in transit to their final location must at all times be protected from frost or drying out with wet sacks or planting bags.

8.9 Ground preparation

Plants must be planted into a medium that has:

1. a texture that retains and releases moisture and nutrients to the plant;
2. a crumb structure that promotes root growth and drainage which prevents waterlogging around the roots.

Ground preparation over an area of at least three times the diameter of the root spread and to a depth of one-and-a-half times the root depth is essential. Heavy clay soils must be made free draining and friable. Light sandy soils need their water-retaining capacity increased. Ground preparation must take place in advance of the planting season when the weather is more reliable, less soil structure damage is caused, and more time is available for planting operations.

8.10 Planting

Final planting can only be carried out when the soil is most friable and not frozen, dry or waterlogged. Before planting, all containers and wrapping must be removed, root-balled plants are to be placed in the pit before the hessian, plastic or other protection to the roots is removed. Damaged and torn roots or stems must be cut back cleanly with a sharp knife or secateurs. Any coiled roots, such as those found in container-grown plants, must be spread out to overcome any future stability problems. Stock should be planted at the same depth as it was grown in the nursery and staked in accordance with one or other of the provisions of BS 4043 and BS 5236 to ensure its stability until established.

Backfilling must be lightly firmed to ensure intimate contact of the soil with the roots. With large plants and trees successive layers of soil will have to be firmed as the backfilling proceeds, to ensure that the plant is securely held but such that the penetration of moisture is not restricted.

8.11 Aftercare

Frost

Frost may loosen the soil around a new plant's roots and this must be remedied as soon as possible by heeling to exclude any consequential air pockets around the roots.

Pruning

Pruning after planting and before leafing out is almost always advantageous to reduce leaf area and the consequent demand on the roots. This may also prevent any die-back that may otherwise occur, and in the long term promotes more vigorous growth.

Weed control

Weed control is essential to provide sufficient moisture and nutrients to the plants until they are well established. This usually takes a minimum of three years and cannot be achieved simply by cutting or mowing surrounding weeds. The application of pre-planting herbicides may help initially but a minimum of one metre (or three times the root spread) diameter area must be maintained weed-free around each tree or shrub until it is fully established. This is usually achieved by mulches, tree spats or herbicides.

Watering

Most plants in south-east and central Britain suffer from a moisture deficiency during the summer months, and all will benefit from irrigation on planting and during their early lives. Any newly planted tree or shrub requires watering after four weeks without an appreciable quantity of rain (less than 30 mm) from mid-May to mid-September. It is therefore prudent to allow on all planting specifications at least 20 litres of water to be provided once every two weeks for standard trees and 10 litres for smaller shrubs. Any smaller quantity will ensure no more than the bare survival of the plant without growth, any larger quantity could be wasteful and leach away the nutrients present in the soil. The overriding principle is to ensure that sufficient water is provided to saturate the full depth of the top soil whether based on clay or sand, and that sufficient waterings are allowed for in tender documents (NBS clause 6401 in Figure 3.1).

Mulching

This cannot only assist in conserving soil moisture but also controls weeds which would otherwise rapidly transpire moisture and evaporation from the soil.

Vermin

Plants can be damaged by vermin such as voles, mice, hares, rabbits, deer and farm stock. Regular inspection, at least once a month, is essential if appropriate action is to be taken in time. In certain circumstances it is best to fence in the whole area; in others, individual plant guards or poison will suffice. (Care must be taken to ensure that when poison is used there is not the slightest risk of harm to livestock, domestic pets or wildlife, for which the contractor will become liable in law.)

Stakes

Stakes and ties must be provided and adjusted regularly to ensure the survival of new planting. It is no longer thought to be always necessary to stake new trees to the underside of the lowest branch. All stakes should be removed as soon as the plant is stable so that the tree may develop its own appropriate stem and root system.

Protective fencing and tree guards

The protection of trees and shrubs must always be measured in detail, and specified accordingly. Such phrases as 'allow for such protection as is thought necessary' are no longer considered adequate. It is incumbent on the specifier in the interest of compatibility of tenders to state precisely where he considers protection to be necessary, and for the tenderer to price accordingly. Suitable clauses are such as those set out in NBS W21 1151–1251 and W31 6051–6651 (part of which is quoted in Figures 3.1 and 3.2).

Vandalism

More trees have died as a result of vandalism than those planted in unsuitable soil conditions. It is essential that those so damaged are replaced immediately and securely protected. Trees should be well staked with adequate tree guards for the full height of the stem.

9

Landscape contract administration

9.1 The landscape architect's administrative duties*

Tendering procedures

The invitation of tenders is a most important activity and one in which the landscape architect has a responsibility to act fairly to landscape contractors in accordance with certain accepted codes and conventions. Those for the building industry are set out in the NJCC Code of Procedure for Single-stage Selective Tendering (Appendix G), and most, if not all, the principles apply equally to the landscape industry. The most important are the following.

1. 'Open' tendering, when tender documents are sent to any contractor in response to a notification in local or technical press is rarely in the best interest of client or contractor and is no longer a practice recommended by the DoE. Where this is still a requirement of standing orders it can best be complied with by investigating the suitability of all applicants and then inviting tenders from the six considered most appropriate. In this context it is only fair to invite tenders from firms of the same size and of the same standards of competence.

   The Code of Procedure for Selective Tendering, agreed with the contractors, the local authorities and DoE, lays down precise requirements for the numbers of contractors invited to tender, the correction of errors and the notification of the results of tenders submitted.

   It is essential to choose the right-sized firms for the particular project and not to expect the highest quality from a low-budget quotation.

* The duties attributable to the 'landscape architect' in this chapter apply equally to landscape managers, supervising officers, and leisure and amenity managers in local authorities. The name 'landscape architect' is not protected as is that of 'architect' by the Architects Registration Act. All that is necessary therefore is for the title of the person to whom these duties are delegated to be inserted in the appropriate place in the Recitals of the JCLI Form of Contract for Landscape Works.

2. Writing or telephoning those selected to inform them of the scope of the project, consultants involved, form of tender documents (i.e. with or without quantities), form of contract and dates for the invitation and submission of tenders, to confirm that they are willing and able to submit a tender by the date and time stated.
3. In the tender documents it is essential, if all tenders are to be submitted on a comparable basis, to state the contract period and whenever possible the anticipated contract commencement and completion dates. If tenders on any other basis are to be considered, i.e. shorter or longer periods, fixed price or fluctuations, then they should be submitted as additional and supplementary tenders. It is then possible, having chosen the lowest tender, to consider and if necessary negotiate the alternative options. It is also essential to state in tender documents the period during which tenders remain open for acceptance (usually 28 days), and which of the two alternatives is to be used for the adjustment of any errors found in the tender prior to the signing of the contract: NJCC alternative 1 (contractors to stand by their tender and absorb any extra costs as a result, or withdraw – commonly known as stick or bust) – or alternative 2 (bona fide errors to be adjusted in the contract sum prior to signing the contract). The former is usually adopted for local authorities and the latter for the private sector. (An example of a Form of Tender is given in Fig. 9.1.)

Dates for the commencement of work on site are of crucial importance and, if not stated in the tender documents, should be ascertained and confirmed prior to their submission. A period of four weeks' notice prior to starting on site is normally essential for the efficient mobilization of the necessary labour by the contractor, and for the procurement of hard landscape and plant materials, particularly those not readily available ex-stock. Should specific items be unavailable it is always recommended that the approval of any substitutes are obtained prior to the submission of tenders and it is always essential to resist the temptation to indulge in or co-operate in 'Dutch' auctions after tenders have been submitted.

If the consultants are not engaged on a full service basis, and post-contract services have been excluded from their responsibility, this should be made clear at the time of tender.

If the period for the execution of the work on site is not included in the tender documents this should be ascertained prior to the acceptance of tenders so that all tenders are submitted on the same basis, regardless of considerations of time for completion of the work. Should any changes subsequently arise these can always be negotiated with the lowest tenderer.

It is also essential to know the date for commencement of the work at the time of tender, since the date for completion of the works has financial consequences if subsequent upkeep or maintenance is required; six months' upkeep during the summer costs considerably more than six winter months.

The NJCC code in respect of the period recommended for the preparation of tenders, the length of time for their acceptance, the adjustment of errors, and the

Tender for Landscape Improvements Church Street Old Peoples Home

To IB Fair ALI
Chief Landscape Architect
Barchester District Council
The Close, Barchester

Sir(s)

I/We having read the conditions of contract specification and schedules of quantity delivered to me/us and having examined the drawings attached thereto do hereby offer to execute and complete in accordance with the conditions of contract the whole of the works described within the period specified from the date of possession for the sum of £13,564
Thirteen thousand, five hundred and sixty four pounds only

I/We agreed that should obvious pricing or errors in arithmetic be discovered before acceptance of this offer in the priced schedules of quantity submitted by me/~~us these errors will be dealt with and adjusted (Alternative 1)~~ will not be adjusted and we will stand by our tender (Alternative 2) in accordance with Section 6 of the current NJCC for Building Code of Procedure for Single Stage Selective Tendering.

This tender remains open for consideration for 28/~~56~~ days (2.) from the date fixed for the submission for lodgement of tenders.

Date this Twenty fifth day of December 19 94

Name J. T. RAMITIN LTD

Address Cold Comfort Nurseries
Barchester Road
Candleford CA5 1XT

Signature B.S.T. Guest
B.S.T. Guest. Chief Estimator

References: 1. Delete as appropriate before issuing.
2. The period specified should not normally exceed 28 days and only in exceptional circumstances should be extended to 56 days.

SUMMARY OF TENDER

Preliminaries	£1500·00
Hard Landscape	5,564·00
Top soil	1180·00
Grassed areas	820·00
Shrubs and Herbaceous Plants	2451·00
Trees	1549·00
TOTAL	13,564·00

Fig. 9.1 An example of a Form of Tender for use with the JCLI Form of Contract.

notification of the results, should be strictly adhered to. Tenders should never be opened as they come in nor left lying around before the deadline for their submission.

When the date and time for the submission for tenders has passed – noon on Monday being commonly used (as postal deliveries can be uncertain, and many tenders are delivered by hand) – it is only fair on those tendering in time to return unopened all those received after the due date marked 'returned – received out of time'. (Proof of date and time of posting are not acceptable. All tenders should be stamped with the date and time of receipt.)

When the tenders have been opened, the landscape architect must list and forward those received to the employer, with his recommendations. This should be followed as soon as possible in a letter or a formal report which in either case should cover the following.

1. The full list of names and tender sums of those received, including those others invited to tender but who did not do so (with the reasons where known). At this stage it is usual to notify all but the lowest two or three that their tender is unlikely to be recommended for acceptance but without any indication of the other tenders received.
2. A financial analysis and arithmetic check on the lowest tender should follow to establish that the tender is fairly, accurately and consistently priced, and free from any major arithmetic errors. Should any errors be found, the tenderer should be given the opportunity to correct or stand by his tender (dependent on which option is stated in the tender document).
3. A recommendation to accept the lowest tender (adjusted where necessary) should follow as soon as possible, with confirmation of his ability to undertake the work within the time laid down and to the standard required. This should be based on the previous experience of the tenderer by the landscape architect, by references taken up prior to the invitation of tenders, or as a result of having worked for the employer before. Under these conditions only rarely will a recommendation for acceptance of a tender other than the lowest be justified.

Only when acceptance of the recommendation has been received should the amounts and names of the other tenders received be notified to the other tenderers (in separate lists, with the amounts in ascending order, followed by the names listed alphabetically, including the lowest. (Figure 9.2 shows an example of a 'standard' letter which can be used for this purpose.) This ensures that, while disclosing the names of the other tenderers and the amounts of the tenders, they cannot be identified with any particular one of them.

Pre-contract meeting

Upon confirmation of the acceptance of the landscape architect's recommendation for the acceptance of a tender a pre-contract meeting should be arranged with the lowest tenderer, at which the following need to be confirmed.

Unsuccessful Tenders

Dear Sir,

Job Reference

....

We acknowledge the receipt of your tender dated ....
but wish to inform you that a more acceptable quotation has been received.

We shall, however, be glad to approach you again for future inquiries.

A list of firms who tendered and a list of the tender prices are attached for your information.

Tenderers (in alphabetical order)

....

....

....

....

Prices

....

....

....

....

Yours faithfully,

Fig. 9.2 An example of a 'standard' letter informing unsuccessful tenderers about other tenders submitted.

1. the names, addresses and telephone numbers of consultants and contractors' staff;
2. dates for possession, and completion, and acceptability of earlier completion dates;
3. distribution and number of copies of contract documents and by whom they are retained;
4. procedure for issuing instructions, variations and certificates;
5. site access, services, signs, security, noise restrictions, working area, etc.;
6. site personnel, working hours, accommodation, safety and welfare requirements;
7. quality control, testing and samples, site works orders, clerk of works, labour returns and weekly reports;

8. dates of regular progress meetings, site inspections and valuations;
9. any other business.

9.2 The contractor's obligations

Under JCLI clause 1.1 the contractor **must always** carry out and complete the works in accordance with the contract documents diligently and in a good and workmanlike manner. **If** the progress of the work is delayed he **must** inform the landscape architect, stating the reasons and estimating the effect on the date for completion (clause 2.2).

9.3 The landscape architect's duties

The landscape architect's duties are no less onerous under JCLI clause 1.2 but more numerous. They are set out in Table 9.1 and include the following.

The landscape architect **must** always supply any further information necessary for the satisfactory completion of the work, issue any instructions necessary, issue interim certificates, certificates of practical completion, making good defects and the final certificate (clauses 1.2, 2.4, 2.5, 2.7, 3.5, 4.2, 4.3 and 4.4).

If he is notified that completion of the works is delayed for reasons beyond the control of the contractor, he **must** extend the contract period by a reasonable amount or recommend the deduction of liquidated or ascertained damages. **If** the works are varied he **must** issue instructions in writing and value the variations in accordance with the contract documents. **If** any errors are found in or between the contract documents he **must** issue the necessary instructions for their correction (clauses 2.2, 2.3, 3.6, 3.7 and 4.1), or issue instructions for the expenditure of any Prime Cost or provisional sums.

If the contractor does not proceed with the work diligently or he goes into liquidation the landscape architect **may** recommend to the employer that the employment of the contractor be determined. He **may** equally instruct that one or any of the contractor's employees is excluded from the works, inspect evidence of insurance, or give instructions for tests and the removal of defective work (clauses 3.2, 3.4, 6.4 and 7.1).

To summarize, the duties of the landscape architect involve:

1. giving general supervision in order to advise the employer on the progress and quality of the work;
2. valuing work in progress, issuing interim certificates for payment of work on account;
3. listing work outstanding at practical completion and defective items at the end of the Defects Liability Period;

Table 9.1 The landscape architect's duties as specified by JCLI

	Must always	Must if need arises	May if need arises
Issue further information necessary, issue certificates, confirm instruction in writing	1.2		
Extend contract period if justified		2.2	
Recommend deduction of damages		2.3	
Certify practical completion	2.4		
Certify when defects made good	2.5 2.7		
Consent to subcontracting			3.2
Exclude contractor's employees			3.4
Issue instructions	3.5		
Issue and value variations including direct losses		3.6	
Instruct expenditure on p.c. & provisional sums		3.7	
Issue instructions to correct inconsistencies		4.1	
Certify progress payments	4.2		
Release retention	4.3		
Issue final certificate	4.4		
Inspect insurance evidence	} client may		6.4
Determine the employment of the contractor			7.1

4. issuing instructions and valuing variations to the contract on the client's behalf, extending the contract period and condemning work not in accordance with the contract;

The landscape architect is paid one fifth of his total fee for post-contract services and the first two occur on every project. The last two items may and often do arise but only when certain events happen, which are usually unforeseen or outside the landscape architect's control.

The employer is responsible to the contractor for honouring certificates which he has agreed (probably without realizing it) that the landscape architect should issue. The landscape architect also has to issue such instructions as are necessary for the completion of the works, and has to grant such extensions to the contract period as are justified. These are matters which the employer puts very low in priority in the list of duties he requires of the landscape architect. Some duties the landscape architect has to do on every contract, some he has to do only if they arise, some he may do if he considers them to be justified.

In addition the landscape architect is responsible to the employer under the landscape architect's appointment for:

1. contract administration;
2. inspecting the progress and quality of work on site;
3. making periodic financial reports.

9.4 Obligatory duties (always necessary)

Landscape architect's instructions (clause 1.2)

A relatively simple requirement provided that they are always in writing, preferably on a readily identifiable form such as that issued by the Landscape Institute (Figure 9.3). Any which involve a material alteration to the authorized expenditure are to be notified to the employer and, preferably, the value agreed with the contractor before it is carried out.

Progress payments (clause 4.3)

These must be certified at intervals of not less than four weeks for the value of work properly executed and any unfixed materials properly but not prematurely brought and stored on site. The value of any work not properly executed and retention must be deducted from the valuation and excluded from the Certificate. (At the same time the client should be notified of any forthcoming variations and increased costs of labour and materials to which the contractor is entitled and which might cause the authorized expenditure to be exceeded.) The Landscape Institute (LI) standard form for Interim/Final Certificates covers these points (Figure 9.4).

Practical completion (clause 2.4)

A matter solely for the opinion of the landscape architect and usually when the works are sufficiently (**not** 'almost') complete to be put to the use intended without risk or inconvenience to the user. A standard form for this has been produced by the LI; alternatively a letter incorporating the following form of words will usually suffice.

I/We certify that in accordance with the conditions of contract, subject to the completion of any outstanding items and/or making good any defects, shortages or other faults which appear during the Defects Liability Period, the Works had achieved, in my/our opinion, practical completion on . . ., the Defects Liability Period will expire on . . ., and that one-half of any retention monies deducted from past previous Certificates are now to be certified for payment.

If only part of the works are to be handed over, these should be identified accordingly together with their approximate value.

The effect of this will be to signify the stopping of any further liquidated ascertained damages which might be deducted, and to notify the employer of his subsequent responsibility for insurance for any subsequent injury to persons or damage to other property and the works.

Landscape Architect's name and address	MANNING CLAMP + PARTNERS 31-32 The Green RICHMOND Surrey	**Landscape Architects Instruction**
Works situate at	Elderly Person Housing Dukes Road, Richmond	
To contractor	Town & Country Landscapes Ltd. Dukes Meadows, Mortlake SW 14	Instruction no. 16 Date

Under the terms of the Contract

dated 3rd January 1985

I/We issue the following instructions. Where applicable the contract sum will be adjusted in accordance with the terms of the relevant Condition.

	Instructions	For office use: Approx costs £ omit	£ add
16.01	PLANTING OMIT: The provisional sum of £500.00 in lieu thereof place your order with Messrs Meadows Nurseries in accordance with their attached quotation dated 9th January 1992 for the surplus of plants in the sum of £978.46 subject to the following i) Substitute Taxus hibernica in lieu Malus £26. 20. ii) Reduce climbers to one of each 41.90 iii) Substitute Forsythia for Hamemelis 70.00 £138.10 iv) 33% reduction in fertiliser and planting 33% x £220 70.00 £208.10 v) Omit watering and replanting 195.86 403.96		
Office reference	Signed ______________ Landscape Architect		

Notes

Amount of contract sum £
± Approximate value of previous instructions £
£
± Approximate value of this instruction £
Approximate adjusted total £

Distribution: Client ☐ Contractor ☐ Quantity surveyor ☐ Clerk of works ☐ File ☐

Fig. 9.4 Interim/Final Certificate. (Reproduced with permission of the Landscape Institute.)

Valuation and Current Financial Statement for Landscape Interim/Final Certificate No.

Employer's name and address: E.O. Burlington Esq. Chiswick House, London W4.

Contractor's name and address: Town and Country Landscapes Ltd. Dukes Meadows Mortlake SW14

Contract: Improvements to Garden Office reference: 17/36

AUTHORISED EXPENDITURE		ESTIMATED FINAL COST	
To amount of Contract	£ 7,500	To amount of Contract	£ 7,500
Additional authorised expenditure	£ 1,250	*Less* contingencies	£ 1,000
			£ 6,500
		Estimated value of Variations to date	+ £ 1,950
			£ 8,450
		Estimated value of Variations yet to be issued	± £ 150
			£ 8,600
Estimated increased costs (if allowable)	£ 750	Estimated increased costs (if allowable)	± £ 750
TOTAL AUTHORISED EXPENDITURE	£ 9,500	ESTIMATED TOTAL COST	£ 9,350

Signed William W. Kent FLI Landscape architect. Date 3.2.91

Distribution: Client ☑ Quantity surveyor ☑ File ☑

Landscape architects name and address: William W. Kent, 12 Carlton House Terrace, London SW1Y 5AH

Landscape Architects Interim/Final

Certificate

Employer's name and address: E. O.Burlington Esq, CHISWICK HOUSE, London W4

Contractor's name and address: Town & Country Landscapes Ltd., Dukes Meadows, Mortlake SW14

Contract: Improvements to Garden

Reference: 17/36

Certificate No.	FIVE
Date of Valuation	3.2.80
Date of Certificate	7.2.80
Valuation of work executed	£ 7,250.00
Value of materials on site	£ 1,250.00
	£ 8,500.00
Less retention	£ 425.00
Total to date	£ 7,075.00
Less amount previously certified	£ 5,000.00
	2,075.00

(Exclusive of any Value Added Tax)

I/We hereby certify that under the terms of the Contract

dated: 3rd February 1991

the sum of (words) Two thousand and seventy five pounds

is due from the Employer to the Contractor

Signed William W. Kent FLI Landscape architect Date: 3.2.91

Distribution: Client ☐ Contractor ☐ Quantity surveyor ☐ File ☐

Fig 9.4 *(Contd)*

Making good any defects (clause 2.5)

Similar wording should be used when the Defects Liability Period has expired, all defective work has been made good, and the balance of retention monies can be released.

Final valuation (clause 4.4)

The landscape contractor is responsible under clause 4.4 of the JCT and JCLI contracts to forward to the Landscape Architect or Quantity Surveyor within three months of Practical Completion (or other such period as has been stated in the tender documents) all the financial evidence he needs to complete his final valuation. As soon as these details have been received, the Defects Liability Period has expired and all defects have been made good, the landscape architect must, within 28 days, issue his Final Certificate releasing the balance of money due to the landscape contractor. Although the final valuation is a matter entirely for the landscape architect or quantity surveyor it is usual good practice (to avoid future disputes, possibly leading to arbitration) to obtain the landscape contractor's written agreement beforehand and to advise the employer of the details prior to the issue of the Certificate.

A brief statement and summary (such as shown in Figures 9.5 and 9.6) will usually suffice.

9.5 Obligatory duties (if they arise)

Approvals (clause 1.1)

It is important to ensure that any 'approvals' of the landscape work – particularly as to the quality of workmanship and materials – are used sparingly if the landscape contractor is not to be relieved of his responsibility to comply with the contract documents in all respects. The same applies to shop drawings which should always be returned with or without 'comment', drawing attention to the need to 'check all setting-out dimensions, tolerances, sections and junctions with adjacent work on site' and never 'approved'.

The landscape architect must, however, always bring to the landscape contractor's attention information of which he is aware (including the route of public service cables) or at least draw his attention to the need to find out!

Extensions to contract period (clause 2.2)

The contract requires that the landscape contractor must notify the landscape architect if completion will be delayed for reasons beyond his control. The

Job..

Contractor..

Statement of final cost

Amount of contract:		£
Less contingencies:		£
Additional works:		
1.		
2.		
3.		
4.		
5.		
6.		
7.		
8.		
9.		
10.		
	£	
Less omissions:		£
1.		
2.		
3.		
4.		
5.		
6.		
	£	£
Net adjustment from adjustment of provisional sums and sundry items:		£
		£

Fig. 9.5 Example of a Statement for Final Valuation.

landscape architect must then extend the contract period for completion of the works by such a period of time as may be reasonable. This calls for a considerable exercise of judgement on the part of the landscape architect, bearing in mind that there must be delays to completion and not just progress); many delays will be concurrent and not consecutive, some being within and some

Job:

Client:

Contractor:

To amount of contract		£
Adjustment of variations prime cost and provisional sums:		£
To total amount of additions:	£	
To total amount of omissions:	£	
Net deduction/extra		£
Total cost of work executed		£

We hereby certify that this Final Valuation properly prepared in accordance with the Terms and Conditions of the Contract is agreed in the above total of
£ (write out in words)
representing the total claim under this Contract.

Signed:

Contractor...

Date..

Architect..

Date..

(Note: this form has to be sent to the contractor in duplicate with a request that he sign and return one copy to the landscape architect.)

Fig. 9.6 Example of a Summary of Account and Final Valuation (for use when no Q.S. is employed).

outside the control of the contractor. The landscape architect must notify the contractor of the extension granted as soon as he is able but is under no obligation to notify him of the reasons or amounts of individual delays reached in coming to his decision.

Only the following are normally considered to be reasons 'beyond the control of the contractor'. They are:

1. *force majeure;*
2. loss or damage occasioned by any one or more of the points referred to in clause 6.3 (loss or damage by fire, storm, flood, etc);
3. civil commotion or strikes affecting the works;
4. compliance with the landscape architect's instructions;
5. delay in the receipt of landscape architect's instructions;
6. delay by persons employed or engaged by the employer not forming part of the contract.

Liquidated and ascertained damages

Should the contract completion be delayed for reasons **not** beyond the control of the contractor and for which the landscape architect is therefore unable to grant an extension he must notify the employer in writing accordingly before the contract date for completion so that the employer deducts the necessary amount of liquidated and ascertained damages from the next Progress Payment.

Variations

All variations which the landscape architect makes must be confirmed in writing and valued by the landscape architect on a fair and reasonable basis, using where relevant the prices in the priced specification/schedules/schedule of rates.

To ensure that these are 'fair and reasonable' the following rules should be followed.

1. Similar work executed under similar conditions and quantity shall be valued in accordance with the rates of the priced specification/schedules/schedule of rates.
2. Similar work executed under dissimilar conditions or quantity shall be in accordance with rates based on those in the priced specification/schedules/schedule of rates/Bills of Quantity.
3. Work not of similar character shall be valued at fair rates and prices with an appropriate adjustment for the cost of preliminary items.
4. Work which cannot be properly valued by measurement shall be valued on the basis of the rules for prime cost of **daywork** as defined by the RICS and BEC at the date of tender subject to the provision of the necessary supporting evidence having been delivered to the landscape architect not later than the end of the week following that in which the work has been executed.

Dayworks can only be used therefore when the variation cannot be 'measured' and if, when acknowledging their receipt, the following words are used, the situation is made clear to all concerned, ensuring that dayworks are only used in the appropriate circumstances.

Thank you for daywork sheets no. . . . Where necessary an Instruction for the variation will follow. This acknowledgement does not, however, indicate our agreement that the Works are variations to the Contract, that if so they will be valued on a dayworks basis, nor that the labour, materials and plant stated are correct.

This is particularly important since the listing of labour and materials used does not necessarily imply that they were necessary for the proper execution of the work. The contractor is only entitled to reimbursement for a reasonable quantity of labour and material. The valuation of work on a daywork basis does not relieve him of his obligation to carry out the work as expeditiously and economically as possible.

9.6 Optional duties (if the landscape architect so decides)

Notice to proceed with the works (clause 7.1)

If it appears to the landscape architect that the landscape contractor is neglecting his primary responsibility to proceed diligently with the works he may, if he thinks fit, suggest to the employer that he, the employer, should exercise his right to determine the contract. If the employer agrees, then a letter should be sent to the contractor by recorded delivery using the following words.

Owing to your neglect and failure to proceed with the above work in accordance with Clause 7.1 of your contract dated . . . WE HEREBY GIVE YOU NOTICE to proceed with the above in a reasonable manner and with due diligence employing the necessary labour and materials for the due performance thereof.
In the event of your neglect to comply with this notice during a period of . . . days from the date hereof, we are instructed by the employer to inform you that he will exercise his powers under Clause 7.1 of the said Contract and determine the Contract.

This usually has the desired effect and either the work is speeded up or the contractor writes setting out the reasons for the delay and the course of action he intends to take to remedy the situation.

Determination of the contractor's employment (clause 7.1)

Should the contractor ignore the foregoing notice then the employer may wish to exercise his right to determine the employment of the contractor. Before so doing, legal advice is essential, with rigid adherence to the proper procedures if, however clear-cut the rights and wrongs of the case may seem to be, the case is not to fail because of such legal technicalities as the service of notice or the contractor being able to establish that he has only 'postponed or suspended' the work.

Exclusion from the works (clause 3.4)

Although this option of the landscape architect to exclude employees of the contractor from the works (no longer dismissed) is clearly set out in the Contract, its formal implementation is rarely necessary. On the infrequent occasions when the need arises, the problem can usually be resolved by a discrete discussion with the landscape contractor without recourse to the legality of the Contract clause.

Expenditure of prime cost and provisional sums, inspection of evidence of insurance, test and/or removal of defective work (clauses 3.5, 3.7 and 6.4)

Should the need for any of these arise, then the necessary provisions of the Contract should be used empowering the landscape architect to issue such instructions as he considers necessary.

Oral instructions (clause 3.5)

Messages must unambiguously indicate to whom they are addressed, whose action is required and if confirmation is necessary. They must be in a form readily understood by the individual concerned.

1. The telephone can be used for simple question-and-answer problems, but not for complicated, difficult or delicate ones. Important points should always be subsequently confirmed in writing.
2. Personal visits are required for all difficult and delicate matters and discussions particularly when personal relationships are involved, and also should be confirmed in writing.
3. Letters and memos must essentially be clear, orderly and concise, and have the advantage that they can be dealt with at the convenience of the receiver.

Periodic inspection

Apart from his duties under the contract, the landscape architect's obligations with his client or employer require him to observe the work to ensure the progress and quality is generally in accordance with the contract documents.

This is part of a separate agreement between the two, and in no way relieves the contractor of his own separate obligation to complete the work in accordance with contract documents.

There is however an implied right of access for the landscape architect to the site in order for him to carry out his other contractual duties under the landscape contract; it is usual for the landscape architect to carry out the necessary inspections in order to discharge his duties to his client at the same time.

Constant inspection

There is no specific provision for a clerk of works in the JCLI Contract although there is no reason why the employer should not employ one.

While the landscape architect is not expected to be on site permanently, but merely to appear at intervals to inspect and check the work, if a clerk of works is employed on the site full-time, he can be expected to ensure conformity with the design although still 'solely as inspector on behalf of the employer' with no power to issue instructions to the contractor. His presence in no way relieves the contractor of his obligations to complete the work in accordance with the contract documents.

A recent case however – following the judgment of *East Ham v. Bernard Sunley* in 1966, before responsibilities for supervision under the 1977 revision to the JCT Form of Building Contract was changed – established that in the event of negligence of the clerk of works his employer was vicariously responsible and the agreed damages were reduced accordingly.

The contractor is responsible under his contract with the employer to carry out and complete the works in accordance with the contract documents. In practice, if he is to submit the lowest tender in competition, this means the lowest standard of workmanship and materials he considers the landscape architect will accept.

Meetings

'Committee' meetings involve a number of people reaching a decision on the basis of a majority vote of its members. This is also binding on the minority who cannot dissociate themselves from the decision without resigning. A chairman conducts the meeting but is not individually responsible for the decisions reached. Other than a few client committee meetings, landscape architects rarely come into contact with this category of meeting.

'Consultant's' meetings when someone meets subordinates (to advise him within their individual sphere of responsibility) so that he can come to a decision within the wider implication of the whole project, are clearly different.

Site and progress meetings are partly 'information' meetings, exchanging and recording information about activities and operations, and partly 'problem-solving' meetings.

Neither consultant's nor site meetings involve a chairman, voting or a corporate responsibility. Matters to be discussed at such meetings should be promulgated beforehand and 'matters arising' kept to a minimum. Minutes should always record decisions without attempting to précis or record discussion, and should clearly indicate what action is required and (in a separate column) by whom. For this reason it is helpful for all concerned if weekly/fortnightly/monthly progress meetings are held at regular intervals in suitable accommodation on site, and a memorandum (not minutes) sets out the decisions in summary form in a format similar to that in Figure 9.7.

9.7 Quality assurance on site

Most mistakes on landscape projects arise through ignorance or carelessness and are rarely deliberate or fraudulent. Site supervisory staff, however, have to devote most of their energies to making sure the necessary labour, plant and materials are available on site when necessary, and when one has been arranged must turn their attention to make sure the next will be present as soon as the first is completed. Any spare time that they have from these tasks is spent in resolving discrepancies in the contract documents and obtaining from the consultants such further details as are necessary to complete the project.

If tenders are obtained in competition the contractor, if he is to submit the lowest, can only include for the minimum standards of workmanship and/or materials to meet the specification.

Although compliance with the contract documents is solely the responsibility of the site agent or 'person in charge' there is clearly a need for periodic spot checks or other procedures to ensure that the quality of the materials specified and included on the drawings and specification are in fact provided.

Landscape architects' Conditions of Engagement with their clients therefore require them to 'visit the site at intervals appropriate to the contractor's programmed activities to inspect the progress and quality of the works. Frequency of the inspection shall be agreed with the client'.

Even if a Clerk of Works has been appointed by the client, or the landscape architect on the client's behalf, to provide constant inspection it is still the landscape architect's responsibility to agree the minimum acceptable standards, initially, for the Clerk of Works, then to ensure that these standards are maintained throughout the rest of the work. It is also important to remember that even the Clerk of Works cannot be in all parts of the site at the same time and the contract documents should therefore provide for predictive inspections of key elements which must not be covered up by subsequent stages of the work before they have been inspected and approved. Such stages usually include:

1. setting out;
2. excavation of topsoil and exposure of formation levels;
3. excavation for foundations;
4. preparation of soil for seeding;
5. tree pits.

Date:

Job title:

Present:	Name	Representing
	....	....
	....	....
	....	....
	....	....
	....	....

Contract sum: £ . . . Certified value to date £ . . .

Contract commencement date ../../.. Contract completion date ../../..

Contract period . . . Weeks contract No . . .

Overall progress . . . % . . .weeks ahead/behind

Total days inclement weather this month . . . Total to date . . .

1.0 Progress review

Item	% Scheduled	% Achieved	Weeks +/− schedule
Progress generally	%..		

2.0 Matters arising from last meeting.

3.0 Construction Problems: Building and civil engineering
Structural engineering
Services engineering
Planting.

4.0 Outstanding information required.

5.0 Quality control/samples

6.0 Any other business.

7.0 Date of next meeting.

Fig. 9.7 Example of an Agenda for a progress meeting.

It is important also to specify 48 hours' notice to be given by the contractor when the previous stage will be ready for inspection if the progress of the work is not to be delayed. Clause 1.1 of the JCLI contract provides for these approvals and thus the landscape architect accepts responsibility for the adequacy of the work and by implication that the work does in fact comply with the contract documents, thereby relieving the contractor of further responsibility. The number of such approvals should be kept to the absolute minimum, and only for items which are difficult to describe or specify accurately such as tree surgery and brick pointing. For these items it is often preferable to restrict in the contract documents the contractor's choice of subcontractors and suppliers to firms of the consultant's own choosing.

The stages at which these approvals can be given will depend on the individual circumstances, some will be immediately after the execution of the work or delivery to site so that the contractor is not unfairly penalized by having to replace at a late stage something which could have been rejected much earlier. Other approvals can only be given after a sufficient period of time has elapsed to establish their efficacy or performance in use over a given period of time, such as the Defects Liability Period.

Some items can **only** be carried out in the presence of the Clerk of Works or landscape architect and these must be made clear in the contract documents. Apart from the tree surgery mentioned earlier this particularly applies to herbicides and fertilizers, the application of which are difficult to confirm after they have been applied without expensive analysis and tests. When such materials have been specified it is essential to establish that they have been provided, as it is not fair to other tenderers, let alone the employer who has paid for it, if an inferior substitute is used or it is omitted entirely by an unscrupulous contractor. Even proof of purchase by delivery notes with the inclusion of the name of the site to which they are to be delivered are rarely an adequate substitute for attendance on site while they are being applied.

BS 5750 1987 – *Quality Systems, Part 2* for use 'where a contract between two parties requires demonstration of a supplier (of goods and services) capability to control the processes that determine the acceptability of the product supplied' might be thought to be the answer to all the problems of site inspection. Unfortunately until recently the adoption of these principles for the management and audit of the quality of building and landscape services has concentrated on the quality of service provided by the firms themselves with certification by a third party and has tended to ignore the problems of the control of quality on site.

The JCLI contract does not require any specific demonstration of the contractor's capability, it merely states the obligation to carry out and complete the works in accordance with the contract documents, leaving it to the landscape architect to establish the contractor's capabilities before including him in the tender list. The practice in other industries to ensure compliance with a specification verified solely by demonstration of capability is quite inadequate for the landscape industry, and the JCLI requirements for 'a competent person in charge to be on the Works at all reasonable times' and the ability 'to exclude any person from the Works' are totally inadequate when the real need is for a management representative free of other commitments whose whole responsibility is to ensure the specified standards are maintained.

The JCLI contract does however prohibit assignment and subcontracting without consent but with no explicit provision for the appointment of a Clerk of Works, for the ability to open up and uncover completed work for inspection, or for the ability to order testing and the removal of defective work. It is essential that these safeguards are included elsewhere in the specification or preliminaries in the bills of quantities.

The provision of an effective quality system for use on site can only be achieved by the landscape contractor providing not only of vouchers to prove the materials and goods are those specified (on a regular basis and not just when asked for) but also a properly structured system of quality records and audit procedures by which the landscape contractor can prove to the landscape architect that by the use of both regular and spot checks he is providing that which has been specified and which he has contracted to provide.

Only by the implementation of such a system can the ever increasing tide of litigation, affecting the building industry and to a lesser degree the landscape industry, be stemmed.

10

Landscape maintenance

10.1 Introduction

Although building owners are usually aware that they are responsible for the regular cleaning, repair, maintenance and servicing of new buildings as soon as they have taken possession and occupied them, e.g. emptying waste paper bins, cleaning floors, windows and lavatories, replacing light bulbs, etc. they rarely, if ever, remember that they are responsible for the same care and attention of the spaces between and around their buildings.

Shrub borders need the almost daily clearing of litter in urban areas, grass needs weekly mowing and watering during the summer months, and even trees need an annual pruning and sometimes tree surgery.

It is clearly inequitable to expect the landscape contractor to include the cost of this work in his initial tender for supply and planting, any more than it is to expect him to provide blanket cover for the cost of replacing any trees and shrubs damaged by vandalism unless it is specified in detail and separately identified. Where this work is required, therefore, it should be clearly itemized in the bills, schedules or tender summary for the landscape contractor to price. Equally, problems will inevitably arise if another subcontractor is employed, or if the employer himself adequately or inadequately accepts the responsibility for carrying out such maintenance. In the event of plants being found to be dead at the end of the Defects Liability Period as a result of substandard stock, inadequate plant handling in transit or negligent planting by the landscape contractor, it is always difficult, if not impossible, to provide sufficient evidence to refute the counter-claim that their death arose from some act or lack of care or inadequate watering on the part of those responsible for their subsequent care during the Defects Liability Period.

It is always essential, therefore, to obtain from the landscape contractor at the time of tender, a separate tender for the regular upkeep of trees, shrubs and grassed areas after practical completion of the work if it is intended that he should be responsible for making good all defects at the end of the Defects Liability Period. In such circumstances, where the landscape contractor is to be held responsible for replacing dead and defective plants and for the upkeep of the landscape works after practical completion and during the Defects Liability Period, this separate agreement should set out in detail the maintenance work to be included in each individual item, for trees, shrubs, herbaceous planting and grassed areas within the site for each season of the year.

Such regular maintenance must be executed with the minimum inconvenience and obstruction of access to the users and occupiers of the site. Any work necessary must be carried out as discreetly and inconspicuously as possible since no work can be tolerated which prohibits the normal access or use of the site. In this connection all sites should provide for storage of landscape equipment in a convenient central location, with hose points at approximately 50-metres centres; equipment and materials do not then always have to be specially brought each time they are needed on site. Landscape areas must also be designed to permit the maximum use of labour-saving equipment, with ramps provided to enable all parts of the site to be accessible (not only to the physically handicapped but also for mechanical equipment) irrespective of the existence of steps elsewhere on site. It is also essential for provision to be made for all machinery, plant and equipment to be returned to the store, all rubbish, weeds and superfluous material to be disposed of and the site to be left clean and tidy on completion.

Any plants found damaged during maintenance operations should be reported to the employer immediately so that remedial measures can be initiated without delay, and all other defects be accepted as the responsibility of the initial landscape contractor.

All maintenance work should be performed at appropriate intervals and in suitable weather conditions for the total number of times stated in the bills, specification or schedules, including any measures specified as being necessary for the application of herbicides and pesticides for the control of disease, weeds and pests.

Most landscape contracts should, therefore, provide for the regular maintenance of planted areas for the first year after practical completion by the contractor who has supplied and planted them. At the same time the contractor then also accepts responsibility for the replacement of all trees, shrubs and grassed areas which may have failed during the same period, for any reason other than theft or vandalism Only in the most rare instances is it advisable for an employer to use his own staff or pay another contractor to carry out this work during the first 12 months after practical completion. The risks of dispute arising from such divided responsibility are self-evident. If the subsequent maintenance is entrusted to the initial contractor problems, arising from lack of communication between the parties resulting in others being blamed for the death of the initial contractor's plants, can be avoided. The work must, however, be clearly described and set out for pricing by the landscape contractor for him to submit an accurate price.

10.2 Tree maintenance

Mature trees require the least frequent regular maintenance of all plants at a negligible cost provided they are inspected at least twice each year to ensure that they remain in a satisfactory and healthy condition. Every May and October, however, roots should be watered with an approved liquid manure (such as 10% nitrogen, 15% phosphoric acid and 10% potash) at the rate 60 grams per square metre. Weeds and grass must on no account be allowed within 500 mm of the stem until such time as the tree is properly established. No evidence as yet exists of the relative advantages of tree spats, mulch, or herbicides. Periodically stakes must be checked, tightened or replaced to ensure that no chaffing of tree bark has been allowed to arise, and any protective vermin spats adjusted. All dead, diseased and broken branches – such as those arising from wind or malicious damage – must be cleanly removed and the scar cleaned up. After a severe frost, the soil around all roots must be firmly trodden down. An experienced arboriculturalist should always be used for pruning so that the natural shape and habit of the tree is preserved. Trees irreparably damaged or destroyed by vandalism must always be replaced as soon as seasonal weather conditions permit.

10.3 Shrub maintenance

Shrubs must be maintained in spring and autumn in the same way as trees. Additionally, at least once each month, during the growing season, unless herbicides have been previously applied, shrub borders should be lightly forked over and any weeds removed, keeping the soil open and friable without disturbing the roots of the plants and the soil around them, then re-firmed.

Live and dead weeds must be cleared away or buried, and perennial weeds must be removed with their roots destroyed. Informal flowering hedges are to be lightly pruned after flowering, to remove dead heads, straggling side and upright shoots but not, under any circumstances, clipped back, or pruned back to a rigid line unless specifically so instructed.

By planting shrubs at comparatively close centres, i.e. 750 mm or less, comparatively weed-free borders are established relatively quickly with subsequent low maintenance costs arising from the minimization of labour expenditure.

10.4 Grass maintenance

Short mown grass is the most demanding of all planted areas in terms of maintenance, requiring to be cut at least once in April, twice in May, three times in June, weekly in July and August, and twice in September plus the inevitable trimming of the grass edges and mowing margins. Allowance must be made also for grass to be rolled at least once in April, June and August.

Long grass, on the other hand, particularly that including bulbs or meadow flowers, must never be cut until late in the summer and that in heavy shade never closer than 75 mm. Fallen leaves must be swept up each October and thereafter as necessary during the winter.

All grass areas must be dressed in spring and autumn with an approved turf fertilizer, selective weed-killer and moss retardant if necessary. Between May and August short-grass cuttings may be collected and spread as mulch over shrub borders, under the ground-cover foliage and forked into the soil in the winter.

10.5 Irrigation

Regular watering must be provided for all newly planted and seeded areas during the first months of the summer if healthy growth of trees, shrubs, herbaceous plants and grassed areas is to be ensured. This must be carried out with a fine rose or sprinkler until the full depth of topsoil is saturated. This is usually necessary once every two weeks in June and September, and weekly in July and August, usually a minimum of twelve waterings a year.

10.6 Planned maintenance

Some planning authorities in granting a consent include a condition laying down the duration during which an agreed landscape scheme must be planted and maintained. The DoE has suggested that maintenance for long periods is unenforceable, implying a continuing obligation on future (unknown) owners. Notwithstanding the requirements of a planning authority, a regular maintenance plan for the building owner to put in hand following the practical completion of a project, is clearly essential.

The various tasks to be carried out on any particular project throughout the year should therefore be scheduled, preferably with guidance as to the estimated time required to carry out each task, so that the actual maintenance costs for each site can be accurately established.

The various seasons' requirements for different planted areas throughout the year make the efficient deployment of labour and machinery throughout the year extremely difficult. Trees require sucker growth and formative pruning only once a year, between June and October. Shrub borders require forking over and spraying with herbicides once a year in April or March. The same border will probably require weeding and hoeing at least six times during the following May-to-September period, and the grass will need to be cut no less than sixteen times during the same period. Table 10.1 summarizes these maintenance requirements.

Table 10.1 Landscape planned maintenance

		Trees			Shrubs			Herbaceous			Meadow grass			Short grass		
	Week No.	Fertilize	Prune	Water	Fertilize/ Weed	Prune	Water	Fertilize/ Weed	Dig	Water	Fertilize/ Weed	Cut	Water	Fertilize/ Weed	Cut	Water
January	1															
	2															
	3															
	4															
February	1															
	2															
	3															
	4															
March	1															
	2	√	√		√	√		√	√		√			√		
	3															
	4															
April	1															
	2				√			√						√	√	
	3															
	4															
May	1														√	
	2				√			√							√	
	3														√	
	4															
June	1						√			√					√	√
	2		√		√			√							√	
	3						√			√					√	√
	4															
July	1						√			√					√	√
	2				√		√	√		√					√	√
	3						√			√					√	√
	4						√			√						√
August	1						√			√		√			√	√
	2				√		√	√		√					√	√
	3						√			√					√	√
	4						√			√						√
September	1						√			√					√	√
	2		√		√			√								
	3						√			√					√	√
	4															
October	1				√			√	√			√			√	
	2															
	3															
	4															
November	1															
	2															
	3															
	4															
December	1															
	2															
	3															
	4															

10.7 Maintenance costs

As can be seen from Table 10.1, a site which may need 10 man-days each month to maintain during the summer may therefore only require less than 3 man-days each month during the winter. Alternative tasks for the surplus 7 man-days each month are therefore essential if the care of the plants is not to be entrusted to casual labour taken on solely for the summer months. This is clearly not fair either to the plants or to the supervisory staff. Activities such as repair and maintenance of hard landscape areas, mechanical plant and grass-cutting equipment are therefore best left to the autumn and winter months when they can be carried out under cover, leaving the replacement of dead and diseased plants to the spring, and keeping the summer free for weeding, watering and grass cutting.

10.8 Time

It must be evident therefore that while the cost of maintaining trees is less than that of shrub borders, and shrubs less than long or short grass, the overriding consideration in the cost of maintaining soft landscape is the economic deployment of the labour involved. If due cognizance is given to the time taken to achieve the various tasks an economic maintenance plan can be established to keep the landscape works in the condition the landscape architect intended.

10.9 Budgeting

It is essential therefore to consider, at the outset of each project, the funds available for subsequent maintenance. High, medium or low maintenance costs are inversely proportional to the initial capital outlay. As a general rule inner urban areas suggest a maintenance-free hard landscape combined with high-maintenance planting, suburban areas an intermediate solution, and for the more rural areas a natural semi-wild low-maintenance solution.

When this has been established a suitable solution can be designed accordingly and the necessary budget for its subsequent upkeep set aside.

10.10 Maintenance contracts

The advent of the 1988 Local Government Act defined maintenance of the ground as:

- cutting and tending grass (including returfing and reseeding but not initial turfing or seeding);
- planting and tending trees, hedges, shrubs, flowers and other plants (but excluding landscaping any area);
- controlling weeds;

and required this to be carried out competitively.

Circular 8/88 issued in the April of that year and 19/88 the following August set out the obligation and procedures for tendering to ensure that even when authorities were in a position to carry out this work themselves they were doing this as economically and efficiently, with rates of return on capital employed and current cost operating surpluses, as if it had been carried out by a private landscape contractor. Although landscape maintenance had been effectively carried out by private contractors for new town and county authorities for a number of years, most local authorities had carried out this work themselves with in-house Direct Service Organisations (DSOs) and were likely to be unsure of specification, tender procedures and supervision of landscape maintenance work to achieve the standards they had previously achieved.

The JCLI in January 1987 therefore published a Model Form of Tender and Contract Documents for Grounds Maintenance, comprising tender preliminaries, specification, Bills of Quantities and contract conditions for this type of work. It is suitable for both lump sum and remeasurement maintenance contracts up to a 1986 value of £150 000 for grass cutting and maintenance of tree and shrub areas including sports grounds (new work is specifically excluded) but allows authorities to select those clauses applicable to each particular project, exclude those inappropriate and add those appropriate to any particular requirement, to ensure that their own tenders and those of others are submitted on a strictly comparable basis.

Where the scope of any particular aspect of the work is uncertain this can be covered by provisional sums or provisional quantities, such as those affecting watering and tree pruning, for subsequent remeasurement and remuneration for the work actually carried out.

The Form of Tender is similar to that adopted by the NJCC for Building for the invitation of single stage selective tenders, and the Specification or Bills of Quantity Preliminary Clauses are based on those prepared by the National Building Specification (NBS) with their guidance notes appropriate for landscape maintenance alongside, the work being carried out under a Landscape Architect, Manager or Authorized Landscape Contract Administrator as appropriate.

The Specification includes regular litter collection, sports field marking, the application of approved herbicides and pesticides, the pruning of trees, hedges and shrubs and the weeding of shrub borders and flower beds.

In some cases the options of performance specification, preferred by some authorities, has been included, e.g. 'cut grass over the specified areas as often as necessary to maintain a length not longer than 50 mm and not shorter than 20 mm'.

The Bills of Quantity include the insertion of day work rates for labour, materials and landscape plant and equipment.

The Conditions of Contract are based on the Standard JCLI Form but with the deletion of the extensions of the contract period damages for non-completion, defects liability retention and malicious damage clauses.

Contracts are envisaged as being for 1–5 years with increased costs evaluated in accordance with PSA Grounds Maintenance Index GM 81, payment being in either 12 equal or unequal monthly amounts or based on the remeasurement of the actual value of work executed during the period in question, whichever basis the authority prefers.

11

Indoor planting

11.1 Introduction

While all the preceding requirements for plants grown externally, with regard to specification, handling and planting, apply equally for indoor plants, there are in addition certain other basic needs which it is essential to comply with if the establishment and survival of indoor plants are also to be ensured.

Plants are unlikely to survive if kept for more than a short while in low light and low temperatures.

11.2 Light requirements

It does not matter if the source is natural or artificial, but no plant can survive in conditions of less than 250 lux, and internal artificial illumination levels greater than 2000 lux, are seldom possible. Some indoor plants may have originated in the tropics or subtropics growing under full sun with light intensities of 1000 lux and others in deep shade of less than 100 lux. However, with gentle acclimatization and a reduction in energy usage many plants when grown indoors will still replace their leaves at 3000 lux or less. Table 11.1 details the light requirements of some common plants and trees.

Supplementary lighting can seldom be achieved from fluorescent 'warm white' lamps, nor with incandescent lighting without a compensating blue filter. High-intensity discharge lamps with a high output in both blue and red wavelengths are available, with an acceptable colour rendering, provided that this is supplied for between 8 and 18 hours per day, 7 days per week but no more and no less.

Table 11.1 Light levels for interior landscape plants. The requirement for individual species is shown as a band on a lux scale meant to represent light available at the foliage canopy. This lux amount is required for 12 to 14 hours a day, 365 days a year.

Key ——— Light level for healthy growth
– – – Minimum level for survival

	250 lux	*500 lux*	*750 lux*	*1250 lux*	*2000 lux*
Trees and large plants (1.5 m to 10 m)					
Araucaria excelsa			–	– – –	———
Cordyline australis			–	– – –	———
Dracaena fragrans massangeana	– – –	———	———		
Dracaena Marginata		– – –	———	———	
Ficus benjamina. P. Elastica			– – ———	———	
Ficus lyrate				———	———
Ficus nitida		– – –	– ———	———	
Kentia forsteriana	– – –	———	———		
Pandanus utilis	– –	———	———		
Phoenix dactylifera. P. canariensis. P. roebelenii			– – –	———	———
Schefflera actinophylla		– –	– – ———	———	
Medium plants (0.5 m to 1.5 m)					
Chamaedora elegans		– – – –	———	———	
Chamaerops humilis				– – ———	———
Codiaeum			– – –	———	———
Cordyline stricta		– – – ———	———	———	
Cordyline terminalis			– –	———	———
Dieffenbachia amoena Tropic Snow		– – – ———	———	———	
Dizygotheca elegantissima				– – – –	———
Dracaena deremensis		– – –	– – ———	———	—
Heptapleurum arboricola		– – –	– – ———	———	
Monstera deliciosa		– – –	– – ———	———	—
Pandanus veitchii			– – ———	———	———
Philodendron Red Emerald, Green Emerald, etc.		– –	– ———	———	
Philodendron pertusum. P. erubescens. P. selloum, etc.		– – – ———	———	———	
Rhapis excelsa			– – – –	———	———
Yucca elephantipes			– –	———	———
Low growing (to 1.0 m) and ground cover					
Bromeliads (*Aechmea, Nidularium Vriesia*, etc.)			– – –	———	———
Aglaonema	– – – –	———	———	———	
Ananas bracteatus		– – –	———	———	———
Asplenium nidus		– – –	———	—	
Chlorophytum variagatum		– – –	———	———	——
Ficus pumila		– – –	———	———	——
Maranta tricolor		– – –	———	———	——
Nephrolepis exaltata		– – –	———	—	
Philodendron tuxla	– – –	———	———	—	
Sanseveiria laurentii		– – –	———	———	——
Scindapsus aurea		– – –	———		

Table 11.1 (*Contd*)

	250 lux	*500 lux*	*750 lux*	*1250 lux*	*2000 lux*
Climbing plants					
Cissus antaratica			- -	———	———
Philodendron scandens. P. pumila		- - -	———	———	
Rhoicissus Ellen Danica		- - -	———	———	———
Plants for cool areas					
Aspidistra elatior	- - - -	———			
Cordyline australis				- - -	———
Fatsia Japonica	- - -	———			
Fatshedera lizei		- - -	———		
Ferns-hardy	———	———			
Medera, green varieties (climbing and trailing)	- - -	———			
Laurus nobilis (bay)		- - -	———	———	———
Ligustrum lucidum		- - -	———	———	———
Pittosporum, various				- - ———	———

Under these circumstances indoor plants will thrive and enhance the interior space for which they are provided.

The various types of lamp available include the following.

1. 'Planting lights'. PAR38 shaped 150-watt colour-corrected incandescent lamps. If fixed 1.0 m above the plant canopy, they will provide 600 lux as flood lights, and 2000 lux as spot lamps.
2. 'Tungsten ballast discharge lamps': provided as fluomeric or blended, they do not require additional control gear. 160-watt lamps will give 300 lux at 1.0 m for floods, and 2000 lux at 2.0 m for spots, having a much longer life than basic 'planting lights'.
3. 'High-intensity discharge lamps': require special control gear to be built in with 250-watt lamps, providing over 3000 lux at 2.0 m.
4. 'Full-spectrum flourescent': the most effective 2 × 1800 mm tubes; at 1.5 m they provide about 1200 lux.

11.3 Acclimatization

Many indoor plants used in Britain are grown in Holland and Belgium, but most larger trees used internally are grown in full sun in Florida or Southern Spain and if they are transported or shipped in total darkness for more than five days can lose 90% of their leaves. While these may be replaced under nursery conditions, they are unlikely to do so if delivered straight to site, since they need time to develop the thinner leaf skin and chlorphyllous structure necessary to take into account their exposure to the reduced light and to develop the necessary greener leaves. The stomata in the underside of the leaves also have to reduce in size in order to take into account the reduced rate of photosynthesis.

The food reserves of the plant will also have been reduced during its transportation to overcome the increased stress to which it has been subjected and to renew at least the foliage lost during the process.

The roots of large trees and shrubs taken from open ground will be immediately subjected to a root/shoot ratio imbalance when placed in the container, which will also take time to readjust.

In general all plants not grown in a glass house or other such controlled environment must therefore have been grown with 55% shade for three months or preferably 70% shade for a minimum of six months prior to having been transferred to the nursery.

11.4 Temperature requirements

Most interior plants thrive in the same temperature conditions as people, i.e. 18–25°C (65–75°F) during the day and 13–18°C (57–65°F) at night with an absolute minimum of 10°C (50°F) for tropical plants and 5°C (40°F) for house plants generally (which is normally achieved in totally enclosed unheated internal spaces) and a maximum of 43°C (110°F) above which photosynthesis stops. Temperatures above or below these limits will inevitably cause damage, especially to indoor plants if they are exposed to them for more than a very short time. The effect of placing a plant directly in the path of an air stream from air-conditioning grilles is particularly dangerous.

Plants such as *Ficus benjamina, Dracaena marginata* and *Howea fosteriana* underplanted with *Spathiphyllum, Mauna loa* and *Philodendum scandens* subjected to temperatures of 20–25°C (68–76.5°F) will recover when heating is reduced to 12–14°C (54–58°F).

Relative humidity will cause greater problems than those of the temperature. Many plants, initially grown in humidities of 60–80%, need time to acclimatize themselves to the 45–55% relative humidity that people normally prefer. Equally, few indoor plants will tolerate a relative humidity below 35%.

It has been suggested that 50% of plant losses are due to incorrect lighting, 40% due to incorrect watering, with only 10% attributable to pests, diseases and other causes. Concentration of effort on the first two will therefore pay dividends.

11.5 Irrigation

Most indoor plants are still dependent on a soil culture with a variety of irrigation systems, although hydroculture is growing in popularity and increased knowledge of the technicalities of its application can often provide markedly better results.

Manual watering

This must be carried out at least once, and sometimes twice, per week. It can, however, over-compact the surface of the soil, restrict the passage of air to the roots and prevent the leaching into them of the nutrients necessary for the plants' survival. Regular trowelling of the surface and proper drainage at the base to dispose of any surplus in manual watering systems is therefore essential. Manual watering is best carried out in heavy doses at well-spaced intervals to allow fluctuation between field capacity and moist state.

Automatic systems

These are available from specialist suppliers to provide soft water on either a drip or a sprinkler basis, but under these circumstances corresponding drainage facilities for the water surplus to the plants' requirements are also essential.

Reservoir-based systems

To overcome the risks of irregular maintenance or to reduce labour costs, indoor plant containers are now often provided with water being fed in from a filler pipe and transferred to the growing medium by capillary action. Such containers, therefore, must be of only such a depth that the water will rise to plants with root balls of a limited size. Such systems, however, can fail if the plants are over-watered.

Hydroculture

The system of growing indoor plants on a container with a clay granule inert supporting medium has been available for many years. The plants (which should be grown specifically for this method for no less than ten weeks after transplanting from compost) then derive their water and food direct from the solution – carefully balanced to provide for the plants' complete needs – in the bottom of the container.

11.6 Soil

All indoor plants should be rooted in uncompacted substrata with at least 20% cavities, and should be salt-free. Root balls must be moist at the time of planting and should be set in a compost such as 6 parts loam, 2 of peat, 2 of bark and 1 part sand without foreign matter or stones over 50 mm diameter. This has been found by experience to suit the majority of plants, provided it contains only a small trace of fertilizer and salt of less than 1000 p.p.m. (the plants having been established in their growing container for at least six months prior to delivery).

The sand must be non-calcareous, coarse-grained, river-washed and free from all extraneous matter. Peat should be coarse-textured, baled, horticultural sphagnum peat to BS 4156.

All compost mixes must completely surround the plant and be measured and blended accurately, thoroughly moistened to between 30–40% with a pH between 5.5 and 6.5. It should be placed in layers no more than 300 mm thick, lightly trodden down and levels made up as and when any settlement occurs.

Soils should be separated from the drainage layer with a synthetic fibre separator to keep the soil fines from entering the drainage layer, which should itself be formed of pre-washed leca placed to a minimum thickness of 100 mm level to within ± 25 mm over any given planter.

11.7 Feeding

Most soil composts contain sufficient fertilizer for at least six months' growth after which time small amounts of liquid feed every 4–6 weeks during the winter, should suffice. Organic fertilizers are not advisable nor are controlled-release fertilizers, unless provided by experts.

With hydroculture, ion exchange resin should be applied every six months with a little liquid feed in soft water areas. Foliar feeds are available but are not generally advisable.

11.8 Maintenance

It is essential that all plants are seen to be in a healthy and vigorous condition when planted, and any not fulfilling this requirement removed from site and immediately replaced by others of the same type and size. A proper maintenance plan – setting out the programme for regular watering, clearing dead plant material and rubbish from containers, feeding, dusting and spraying with pesticides – should be immediately put in hand and continued for a period of at least 12–36 months.

Dust can be initially removed by wiping with a damp cloth or feather duster; heavier deposits require the use of a mild detergent, applied with a fine spray using soft water. One or two monthly treatments of white spirit or 'leaf shine' can sometimes be used for some dicotyledons to reduce dust but is not advisable except in special circumstances and then only under close supervision.

An Agreement made this ______ day of ______ One thousand nine hundred and ______ between ______ whose office is situate at ______ (hereinafter called "the Company" which expression includes its agents or servants) of the one part and ______ (hereinafter called "the Owners") of the other part.

Whereby it is agreed as follows:

1) The Company shall from the ______ day of ______ One thousand nine hundred and ______ until the ______ day of ______ One thousand nine hundred and ______ keep the displays of plants described in the Schedule hereto (hereafter referred to as "the plants") full and the plants clean and cared for and will replace dead or dying plants with others of a like nature. For this purpose the Company shall send a representative once every ______ weeks to inspect and attend to the plantings in accordance with the terms hereof.

2) This contract shall remain in force for twelve months from the above date and thereafter be renewed annually until cancelled by either party in the manner set forth in paragraph 5 hereof.

3) In consideration of the above service performed or about to be performed the Owners will pay to the Company the sum of £______ per calendar month plus VAT to be invoiced on a three monthly basis in advance. This sum to be subject to an annual revision equal to the inflation rate notified within the Retail Price Index in existence during the last month prior to renewal date.

4) This contract does not include accident or misuse of plants by the Owners, their employees or third parties and is made on the understanding that the Owners will continue to supply the environmental conditions in existence when the contract is first signed. In the event of plant loss being caused by temperatures falling below 10 degrees centigrade/50 degrees fahrenheit replacement plants will be charged for unless a colder temperature regime has been agreed in writing. This also applies to displays being moved to unsuitable positions without the Company's consent and in such cases any untoward plant replacement will be covered by an additional charge. Water will also be provided to the Company free of charge by the Owners.

5) Beyond the primary period of contract, i.e. twelve months, the agreement shall continue on the same terms and conditions on an annual basis subject to the consent of the Owners and subject to the right of the Owners to cancel the agreement subject to three months' notice in writing prior to the end of the second or subsequent years.

Schedule of items covered by this Contract

The Owners ______	The Company ______
______	______
Signed by ______	Signed by ______
Title ______	Title ______
Date ______	Date ______

Fig. 11.1 BALI maintenance contract form, for interior plant displays.

A suitable maintenance contract is given in Figure 11.1 which is based on the standard BALI maintenance contract for interior plant displays. (Plant losses of 10% per annum must be expected and allowed for.)

11.9 Pests and diseases

Although indoor plants are usually supplied relatively free from pests, they rapidly lose their resistance, due to the low light levels.

Tree cuts should be treated with a fungicidal sealant and the indoor plant maintenance contractor must allow for all necessary spraying or dusting against pests and diseases. Fungicides, herbicides and insecticides must be non-toxic to normal use by people, birds and animals, and be included in the Agricultural Chemical Approval Scheme's list of products. It is essential that they are always applied strictly in accordance with the manufacture's printed instructions, whether as contact or systemic compounds. Pyrethrum, resmethrin, melathin, kelthane and methylated spirits are the most effective and most popular contact herbicides for aerial pests and diazinon for subterranean ones.

Appendices

A Summaries of landscape British Standards
B Landscape subcontract
C JCLI Standard Form of Agreement for Landscape Works
D JCLI Supplement to the JCT Intermediate Form IFC 84
E JLCPS General Conditions, Specifications and Schedules of Quantity for the Supply and Delivery of Plants
F JCT Standard Form of Tender by Nominated Supplier (Tender TNS/1)
G NJCC Code of Procedure for Single-stage Selective Tendering

Appendix A Summaries of landscape British Standards

BS 1192: –
Construction drawing practice

BS 1192: Part 1: 1984 ≠ISO 6284
Recommendations for general principles
Sets out general principles for the preparation of all drawings for the construction industry, including schedules prepared on drawing sheets and numbered as one of a drawing set. Applicable to drawings produced by computer techniques, and to drawings that will be reproduced by microfilming in accordance with BS 5536.
36 pages Gr 8 BDB/4
0 580 13651 5

BS 1192: Part 2: 1987
Recommendations for architectural and engineering drawings
Illustrates the general principles given in BS 1191: Part 1, and the application of the symbols and conventions given in BS 1192: Part 3.
60 pages Gr 9 BDB/4
0 580 15890 X

BS 1192: Part 3: 1987 = ISO 4067/2, ≠ISO 4067/1. ISO 4067/6, ISO/TR 8545
Recommendations for symbols and other graphic conventions
Does not include symbols applicable to fire protection drawings and to signs for use on buildings or equipment.
Replaces PD 6479: 1976.
AMD 5882, February 1989
52 pages Gr 8 BDB/4
0 580 15896 9

BS 1192PG: 1988
Graphic symbols for construction drawings
A pocket guide derived from BS 1192: Part 3: 1987. It is not itself a British Standard. Gives all the symbols in the Standard except those for land use.
Also numbered IP 001: 1988.
21 cm × 10 cm, foldout card Gr 2 *Special prices are available for orders of more than 9 copies.*

BS 1192: Part 4: 1984
Recommendations for landscape drawings
Recommendations for the preparation of landscape drawings and schedules. Includes symbols and abbreviations which are used in a series of typical drawings. Appendices include summaries of information commonly used when landscape drawings are being prepared.
40 pages Gr 8 BDB/4
0 580 13652 3

BS 1192: Part 5: 1990 ≠ISO/TR 10127
Guide for structuring of computer graphic information
Gives guidance on the representation of construction by computers, primarily for the purpose of generating drawings and the exchange of data between CAD users.
24 pages Gr 7 BDB/4
0 580 18277 0

BS 1830
Not allocated.

BS 1831: 1969 ≠ISO 765, ISO 1750
Recommended common names for pesticides
Coined common names for 278 pesticides. Chemical names, structural formulae, classification according to use (acaricide, avicide, bactericide, fungicide, herbicide, insecticide, molluscicide, nematicide, plant growth regulator, rodenticide) and other non-proprietary names. Common names for pesticides of uncertain composition, common names considered by ISO but not recommended, organic pesticides not requiring common names. Index of common names, chemical names, other names listed.
Further approved names are listed in special announcements in BSI News. Replaces Supplement No. 1 (1967) to BS 1831: 1965. Partially replaced by BS 1831: Part 1: 1985.
AMD 3097, July 1979 (Gr 0)
AMD 4597, July 1984 (Gr 2)
A5, 108 pages Gr 8 CIC/48
0 580 05349 0

Supplement No. 1 (1970) to BS 1831: 1969
Provides a further 30 recommended common names for pesticides.
AMD 1774, June 1975 (Gr 0)
AMD 4598, July 1984 (Gr 0)
A5, 12 pages Gr 3 CIC/48
0 580 06165 5

Supplement No. 2 (1970) to BS 1831: 1969
Provides a further 32 recommended common names.
AMD 4175, February 1983 (Gr 0)
AMD 4599, July 1984 (Gr 0)

A5, 12 pages Gr 3
0 580 06165 5 CIC/48

Supplement No. 3 (1974) to
BS 1831: 1969 ≠ISO 1750
A further 119 recommended common names.
AMD 1774, June 1975 (Gr 0)
AMD 4600, July 1984 (Gr 1)
A5, 44 pages Gr 7 CIC/48
0 580 08830 8

BS 1831: –
Common names for pesticides

BS 1831: Part 1: 1985 –ISO 257
Guide to principles for selection of common names
Definition, purpose, principles for selection, recommended syllables and style of writing and printing. Appendices give procedure for international establishment of common names, corresponding UK procedure and key to pronunciation.
Partially replaces BS 1831: 1969.
16 pages Gr 6 CIC/48
0 580 14763 0

BS 2468: 1963
Glossary of terms relating to agricultural machinery and implements
Contains terms and definitions relating to agricultural machinery, implements, parts and accessories, covering a wide range of agricultural work. Terms are grouped under appropriate classes and sections and a comprehensive alphabetical index is provided. Numerous illustrations are also included in the glossary.
AMD 4395, October 1983 (Gr 0)
A5, 96 pages Gr 8 AGE/11

BS 3882: 1994
Recommendations and classification for top soil
Description of top soil; classification by texture; classification by soil reaction (pH); classification by stone content; size of stones. Notes on method of test for top soil; designation of top soil.
AMD 3089, May 1979[R]
A5, 12 pages Gr 3 AFC/20

BS 3936:—
Nursery stock

BS 3936: Part 1: 1992
Specification for trees and shrubs
Conifers and woody climbing plants, suitable to be transplanted and grown for ornament or amenity. Root systems, condition, dimensions, labelling, forms and dimensions. Examples of plants appended for information.
24 pages Gr 7 AFC/1
0 580 20234 8

BS 3936: Part 2: 1990
Specification for roses
Requirements for rose plants propagated by any method. Excludes roses grown purely for indoor use.
AMD 6628, September 1990
8 pages Gr 3 AFC/1
0 580 17757 2

BS 3936: Part 3: 1990
Specification for fruit plants
Also gives details for specific plants.
12 pages Gr 5 AFC/1
0 580 18103 0

BS 3936: Part 4: 1984 (1989)
Specification for forest trees
Requirements for origin, dimensions, root and shoot, description, condition and packaging of forest trees for timber production.
8 pages Gr 3 AFC/1
0 580 13945 X

BS 3936: Part 5: 1985
Specification for poplars and willows
Specifies requirements for dimensions, shoot condition and marking for plants to be grown for timber production, shelter belts or for amenity.
8 pages Gr 3 AFC/1
0 580 14747 9

BS 3936: Part 7: 1989
Specification for bedding plants
Applies to the most common ranges of ornamental plants and vegetables grown and sold for planting out directly into open ground.
8 pages Gr 3 AFC/1
0 580 17070 5

BS 3936: Part 9: 1987
Specification for bulbs, corms and tubers
Dimensions, condition, time of flowering and labelling of the more commonly grown plants.
8 pages Gr 3 AFC/1
0 580 16271 0

BS 3936: Part 10: 1989
Specification for ground cover plants
Root system, condition, dimensions, designation and labelling with information on whether plants are deciduous, evergreen or herbaceous.
10 pages Gr 5 AFC/1
0 580 17758 0

BS 3936: Part 11: 1984 (1989)
Specification for container-grown culinary herbs
Definition, requirements for root-system, dimensions, designation and labelling. Appendix gives common and botanical names of 30 plants and indicates whether they are annual, biennial or perennial and hardy or half-hardy.
AMD 6256, April 1990
4 pages Gr 2 AFC/1
0 580 13497 0

BS 3969: 1990
Recommendations for turf for general purposes
Characteristics and handling of various types of turf for use in different situations.
16 pages Gr 6 AFC/20
0 580 17509 X

BS 3998: 1989
Recommendations for tree work
Recommendations for the planning and undertaking of work relating to safety, sterilization of tools, tree wounds, decay and pruning, bracing, tree nutrition and repair work to roots and bark.
AMD 6549, December 1990
20 pages Gr 7 BDB/5
0 580 17170 1

BS 4043: 1989
Recommendations for transplanting root-balled trees
Recommends techniques for transplanting trees, including those which have been prepared for transplanting and those which have not.
Replaces BS 5236: 1975.
16 pages Gr 6 BDB/5
0 580 17144 2

BS 4156: 1990
Recommendations for peat for horticultural and landscape use
Gives main characteristics of peat with methods of test and information on uses.
16 pages Gr 7 AFC/20
0 580 18056 5

BS 4428: 1989
Code of practice for general landscape operations (excluding hard surfaces)
Preliminary investigations, drainage, grading and cultivation, seeding of grass areas, turfing, tree planting, woodland planting, planting of shrubs, herbaceous plants and bulbs.
AMD 6784, September 1991 (Gr 1)
36 pages Gr 8 BDB/5
0 580 17194 9

BS 5551:—
Fertilizers

BS 5551: Part 1:—
Terminology and labelling

BS 5551: Part 2:—
Sampling

BS 5551: Part 3:—
Physical properties

BS 5551: Part 4: Section 4.1:—
Determination of nitrogen

BS 5837: 1991
Guide for trees in relation to construction
Principles to be applied to achieve a satisfactory juxtaposition of trees, including shrubs, hedges and structures. It follows, in sequence, the stages of planning and implementing the provisions which are essential to allow developments to be integrated with trees.
38 pages Gr 8 BDB/5
0 580 20245 3

BS 7370:—
Grounds maintenance

BS 7370: Part 1: 1991
Recommendations for establishing and managing grounds maintenance organizations and for design considerations related to maintenance
Gives guidance on landscape planning.
60 pages Gr 8 EPC/2
0 580 19008 0

BS 7370: Part 3: 1991
Recommendations for maintenance of amenity and functional turf (other than sports turf)
Categories of turf (other than sports turf), recommending maintenance objectives.
76 pages Gr 8 EPC/2
0 580 19009 9

Appendix B Landscape subcontract

Form of Tender and Conditions of Subcontract

For use with the JCLI Agreement for Landscape Works.

Information for Tenderers

1. Contract name and location: ....

2. Employer: ....

3. Contractor: ....

4. Landscape Architect: ....

5. The Form of Contract is the JCLI Agreement for Landscape Works, completed as follows:

   Clause 2·1: The Works may be commenced on .... and shall be completed by ....

   Clause 2·3: Liquidated damages: £ .... per week or part of a week.

   Clause 2·5: Defects Liability Period: .... months

   Clause 2·7: Plant Failures grass .... months after Practical Completion

   shrubs and trees .... months after Practical Completion

   semi mature and ELNS trees .... months after Practical Completion

   Clause 4·4 Documents for Final Valuation ....

   Clause 4·5 Contributions levy and tax changes ....

   Clause 6·3 A/B Insurance against fire .... clause A/B deleted.

   Clause 6·3A Percentage to cover professional fees ....

   Clause 6·5B Provisional sum for malicious damage ....

   *Clause 6·3A will be deleted.
   *Clause 6·3B will be deleted, the insurance under clause 6·3A being the full value plus ....%

6. The Subcontractor shall, before tendering, visit the Site or otherwise make himself familiar with the extent of the Subcontract Works, the Site Conditions, and all local conditions and restrictions likely to affect the execution of the work. The Subcontractor may have access to any available drawings or work programmes relative to the preparation of his tender on request and by appointment.

7. Two copies of this Form are provided – One copy to be priced, signed by the Subcontractor and returned to .... the other is for his retention.

Form of Tender

1. Tender price for ....
(The Enquirer should enter trade or brief description of Works)

.... Total tender price

The Tenderer should enter the amount of his tender in words and figures.

2a The tender is to be adjusted for price fluctuations of labour only/materials only*/labour and materials* as set out in Subcontract Condition 4·4, a list of relevant basic prices being enclosed.

2b The tender is to be adjusted for price fluctuations by use of the NEDO Price Adjustment Formula Indices (Category ....)/the Specialist Engineering Installation Indices (Category ....)*.

*2c The tender is to be on a firm price basis.

3. **Schedule of Daywork Charges**
The Subcontractor is requested to insert his hourly daywork rates in the space provided in the Schedule below. His rates for labour shall be deemed to include overheads and profit and all payments in connection with Holidays with Pay, Bonus and Pension Schemes, Subsistence Allowances, Fares and Travelling Time, Imported Labour Costs, Non-Productive Overtime Costs, and any other payments made under the Working Rule Agreement, any Regulation, Bye-law or Act of Parliament. The Subcontractor is also invited to insert in the space provided the percentage addition he will require for his overheads and profits on the nett cost of materials and on plant charges.

(i) **Labour**

Craftsmen @ ....£ .... per hour

Labourers/Mates @ .... £ .... per hour

(ii) **Materials and Plant**
Materials invoice cost plus .... %

Plant Charges plus .... %

4. Period of notice required before commencing the Subcontract Works: ....
Time required to complete the Subcontract Works (the Subcontract Period unless agreed otherwise): ....

5. We, the undersigned, agree that this quotation will be open for acceptance within .... week(s) and that we have read and understand the terms and conditions printed overleaf and that should our quotation be accepted we will enter into a Subcontract in accordance with the said terms and conditions.

For and on behalf of ....

....

Signed .... Date ....

*Delete if not required.

Conditions of Subcontract

1·0 Intentions

Subcontractor's Obligation

1·1 The Subcontractor shall with due diligence and in a good and workmanlike manner carry out and complete the Works in accordance with the Subcontract Documents using materials and workmanship of the quality and standards therein specified provided that where and to the extent that approval of the quality of materials or of the standards of workmanship is a matter for the opinion of the Landscape Architect such quality and standards shall be to the reasonable satisfaction of the Landscape Architect.
No approval expressed or implied by the Contractor and/or the Landscape Architect shall in any way relieve the Subcontractor of his responsibility for complying with the requirements of this Subcontract.

Principal Contract and Special Conditions

1·2 The Subcontractor is deemed to have full knowledge of, and so far as they are applicable to the Works agrees to comply with, the provisions of the Principal Contract as though the same were incorporated herein and the Main Contractor were the Employer and the Subcontractor were the Contractor. Any conditions contained in the Subcontractor's Tender shall be excluded.

Information provided by others

1·3 The Subcontractor must make written application to the Contractor for instructions, drawings, levels or other information at a date which is not unreasonably distant from nor unreasonably close to the date on which it is necessary for the Subcontractor to receive the same.

Information provided for others

1·4 Any instructions, drawings, levels or other information relating to the Works which is requested from the Subcontractor must be provided in due time and so as not to cause disruption or delay to the works to be performed under the Principal Contract.

2·0 Commencement and Completion

Progress and Completion

2·1 The Works are to be commenced within the period of notice stated on the Form of Tender and are to be completed within the Subcontract period subject only to such fair and reasonable extension of time as the Contractor shall allow. The Works are to be carried out diligently and in such order, manner and time as the Contractor may reasonably direct so as to ensure completion of the Principal Contract Works or any portion thereof by the completion date or such extended date as may be allowed under the Principal Contract. If the Subcontractor is in breach of the foregoing he shall pay or allow to the Contractor the amount of loss or damage suffered by the Contractor in consequence thereof.

Overtime

2·2 No overtime is to be worked without the Subcontractor first obtaining the consent in writing of the Contractor. No additional payment for overtime will be made unless the Subcontractor is so advised in writing by the Contractor and, if the Subcontractor is so advised, he will be reimbursed the net additional non-productive rate incurred, including any net additional cost of Employers' Liability and Third Party Insurances. The Subcontractor will be required to obtain any necessary overtime permit from the appropriate authority.

Annual Holidays

2·3 Under the Annual Holiday Agreement, the Site will be closed down for certain periods which may be whilst the Subcontractor's work is in progress. The Subcontractor will be deemed to have included in his Tender for any additional costs and time resulting from such closure.

Maintenance and Defects Liability

2·4 The Subcontractor will (1) maintain the Works at his own expense to the Contractor's and the Landscape Architect satisfaction both during the progress of the Works and until the Landscape Architect has issued a Certificate of Practical Completion including the Works and (2) make good at his own expense, and at a time to be decided by the Contractor, any defects or damage to the Works.

3·0 Control of the Works

Assignment

3·1 Neither the Contractor nor the Subcontractor shall, without the written consent of the other, assign this Contract.

Use of Site

3·2 The Site shall not be used for any purpose other than for the carrying out of the Works. Works to be executed outside the Main Contractor's Site boundary shall be carried out to suit the convenience of adjacent occupiers or Local Authorities at times to be agreed by the Contractor in writing.

Variations

3·3 No variation shall vitiate this Subcontract. The Subcontractor shall advise the Contractor in writing of all work involving a variation or extra work within 14 days of such variation or extra work becoming apparent, at the same time submiting detailed and priced calculations based upon this Subcontract showing such price adjustment, if any. Variations or extra work shall not be undertaken by the Subcontractor nor shall he receive payment for such variation or extra works without written authority from the Contractor.
Where variations or extra works cannot be valued by reference to this Subcontract then the value of such variations or extra works shall be subject to agreement between the Contractor and/or the Landscape Architect and the Subcontractor.

Dayworks

3·4 No daywork will be permitted except where in the opinion of the Contractor and/or the Landscape Architect, it would be unfair to value such work at other than daywork rates. Where the Subcontractor considers he has claim to daywork due notice must be given and valuation by daywork approved by the Contractor in writing prior to the execution of the work in question in order to facilitate checking the time and materials expended thereon. All daywork sheets shall be rendered by the end of the week during which the work is executed. All daywork will be paid for at the rates stated on the Form of Tender.

Adjustment for Provisional Sums

3·5 Instructions will be issued in respect of Provisional Sums. No loss of profit will be allowed in respect of such instructions.

4·0 Payments

Discount to the Contractor

4·1 The Subcontractor will allow for all payments to be made in full within 17 days of the date of the Landscape Architect's Certificate to the Employer without any cash discount for prompt payment. Any such discount included on the Form of Tender will be deducted from the tender sum before the order for the subcontract works is placed.

Progress Payments

4·2 Payment will, subject always to these terms and conditions, be made to the Subcontractor as and when the value of such Works under the terms of the Principal Contract is included in a Certificate to the Contractor and the Contractor receives the monies due thereunder. Applications for payment are to be rendered to the Contractor in duplicate by the Subcontractor.
Payment shall be by instalments of the rate of:
95% of the value executed as the Works proceed.
2½% upon practical complete of the Works.
2½% on satisfactory completion of making good defects under the Principal Contract or as soon as the final account for all Works executed under this Subcontract shall have been agreed, whichever may last happen.
Progress payments shall be on account only and shall not be held to signify approval by the Contractor and/or the Landscape Architect of the whole or any part of the Works executed nor shall any final payment prejudice any claim the Contractor may have in respect of any defects in the Works whenever such defects may appear.

Estimates of Loss, etc.

4·3 In addition to the Contractor' Common Law rights of set off, if the Subcontractor shall cause the Contractor loss by reason of any breach of this Contract or by any tortious act or by any breach of statutory duty giving rise to a claim for damages or indemnity or contribution by the Contractor against the Subcontractor, or the Contractor shall become entitled to payment from the Subcontractor under this Contract, then without prejudice to and pending the final determination or agreement between the parties, the Contractor shall bona fide estimate the amount of such loss, indemnity or contribution or payment, such estimate to be binding and conclusive upon the Subcontractor until such final determination or agreement.

Fluctuations (to apply only if item 2a of the Form of Tender is completed).

4·4 The sum or sums referred to in this Subcontract shall be based upon the rates of wages and such other emoluments, allowances and expenses (including the cost of Employers' Liability and Third Party Insurances) as are properly payable by the Subcontractor to work-people engaged upon or in connection with the Works in acccordance with the rules or decisions of the wage fixing body of the trade or trades concerned applicable to the Works. Such rates of wages and the prices of materials shall be as detailed in the Basic Price List as provided by the Subcontractor and attached hereto.
Should any fluctuations from the Basic Price List occur during the currency of this Subcontract, the net additional cost actually and properly incurred or saving that ought to have been made, by such fluctuations shall be added to or be deducted from the total amount payable under the terms and conditions of this Subcontract. Fluctuations in the cost of materials will be adjusted net.
Provided always that immediate notice in writing shall be given of such fluctuations, and an approved weekly return submitted to the Contractor showing the total number of men and hours and the deliveries of materials effected for detailed checking by the Contractor and Landscape Architect.

5·0 Statutory Obligations:

Safety, Health and Welfare

5·1 The Subcontractor shall comply with the Contractor's requirements on matters affecting the safe conduct of work on the Site and all statutes, bye laws and regulations affecting the Works and the carrying out thereof.

Statutory Payments

5·2 The Subcontractor shall include in his quotation for any payments to be made under the Working Rule Agreement, all payments in connection with holidays with Pay, Bonus and Pension Schemes, National Insurance, Subsistence Allowances, Fares and Travelling Time, Imported Labour Costs or any payments required by Regulations, Bye-law or Act of Parliament.

6·0 Injury, Damage and Insurance

Responsibilities of the Subcontractor

6·1 The Subcontractor shall indemnify the Contractor against all claims, causes of action, costs, loss and expense whatsoever in respect of:

1. Personal injury or death of any person or injury or damage to any property real or personal arising out of or in the course of or caused by any works executed by the Subcontractor and/or the execution of such works (including but not restricted to the use of any plant, equipment or facilities whether in connection with such execution or otherwise) and/or any design undertaken by the Subcontractor and
2. Any negligence or breach of duty on the part of the Subcontractor, his Subcontractors, his or their servants or agents and
3. Any breach or non-performance or non-observance by the Subcontractor, his Subcontractors, his or their servants or agents of the provisions of the Principal Contract in so far as they relate or apply to the Works and are not inconsistent with the provisions of this Subcontract.
4. Any act, omission, default or neglect of the Subcontractor, his Subcontractors, his or their servants or agents which involves the Contractor in any liability under the Principal Contract.
5. Any damage, claim loss or expense to or involving any plant (whether of the type aforesaid or otherwise) hired or loaned or otherwise made available to the Subcontractor or operating for the Subcontractor's benefit.

Responsibilities of Others

6·2 The Subcontractor shall not be responsible for loss or damage caused by fire, storm, tempest, lightning, flood, bursting and overflowing of water tanks, apparatus or pipes, earthquake, aircraft or anything dropped therefrom, aerial objects, riot and civil commotion, to the Works or to any materials (other than temporary buildings, plant, tools, scaffolding and machinery provided by the Subcontractor, or any scaffolding or other plant which is loaned to him by the Contractor), properly upon the Site and in connection with and for the purpose of the Subcontract. In the event of any such loss or damage, the Subcontractor shall, if and when directed by the Contractor in writing, proceed immediately with the rectification or replacement of the damaged work and materials and the erection and completion of the Works in full accordance with the terms, provisions and conditions hereof, and expenses in respect of any of the matters referred to in sub-clause 6.1.1 and 6.1.2 above and shall on demand produce to the Contractor adequate evidence of such insurance.

Subcontractor's Work, Materials and Plant

6·3 The Works, materials, tools, plant, scaffolding, machinery and buildings of the Subcontractor, the subject of or used in connection with this Subcontract whether on Site or not, shall in every respect be at the Subcontractor's risk (except those risks for which the Subcontractor is not responsible under Clause 6·2).

Subcontractor's Insurance

6·4 The Subcontractor shall adequately insure:

1 His and the Contractor's liability in respect of any claims, causes of action, costs, losses and expenses in respect of any of the matters referred to in sub-clauses 6·1·1 and 6·1·2.

2 Against all Employers' Liability and Third Party (including Third Party Fire) risks arising out of the execution of the Works.

The Subcontractor shall produce on demand policies of such insurances, together with receipts for premiums, or other adequate evidence of such insurance.

In case of neglect by the Subcontractor to effect the Insurances, the Contractor shall be at liberty to insure on behalf of the Subcontractor and to deduct the premium so paid from any monies due or becoming due to the Subcontractor.

7·0 Determination

Determination by the Contractor

7·1 The Contractor may without prejudice to any other of his rights or remedies determine the Subcontractor's employment under this Subcontract if the Subcontractor:-

1. fails forthwith upon notice from the Contractor to commence remedial work to any defective workmanship and/or materials or fails to proceed with the same with due diligence or to complete such remedial work to the satisfaction of the Contractor or the Landscape Architect within a set period as the Contractor may specify in the said notice or if none is so specified within a reasonable time.
2. fails to withdraw immediately, at the request of the Contractor, any one or more of his employees to whom the Contractor objects or whose presence on the Works may contravene the conditions of this or the Principal Contract, or may cause labour disputes in the Subcontractor's or any other trade, and to replace such employees within a reasonable time by others against whom there is no such objection.
3. makes any arrangements with his creditors, has a Receiving Order made against him, executes a Bill of Sale, or commits an act of bankruptcy or, being a limited company, goes into liquidation, or has a Receiver appointed.
4. fails within seven days' notice in writing from the Contractor to comply with any of the obligations on the part of the Subcontractor herein contained.

Upon determination by the Contractor the Subcontractor shall not remove any of his equipment, materials or property from the Site and, notwithstanding anything contained in these conditions, shall be entitled to no further payment until completion of the Works by the Contractor or by others whereupon the Subcontractor shall become entitled to payment for Works executed and materials provided by the Subcontractor subject always to the right of the Contractor to set off all losses expense and damages suffered or which may be suffered by the Contractor by reason of such determination and subject further to any other right of set off which the Contractor may have. For the purposes of such completion the Contractor shall have the right to use the Subcontractor's equipment, materials and property on the Site and to any materials or fabricated work lying at the Subcontractor's works or workshop which have been bought or fabricated for the purpose of this Subcontract.

Determination by the Subcontractor

7·2 The Subcontractor may without prejudice to any other of his rights or remedies determine the Subcontractor's employment under this Subcontract if the Contractor:

1. fails to make any payments in accordance with this subcontract.
2. unreasonably attempts or obstructs the carrying out of the Subcontractor's Works
3. makes any arrangements with his creditors, has a Receiving Order made against him, executes a Bill of Sale, or commits an act of bankruptcy or, being a limited company, goes into liquidation, or has a Receiver appointed.

Upon determination by the Subcontractor the Contractor shall pay to the Subcontractor, after taking into account amounts previously paid, such sum as shall be fair and reasonable for the value of work begun and executed, materials on site and the removal of all temporary buildings, plant tools and equipment. Provided always that the right of determination shall be without prejudice to any other rights or remedies which the Subcontractor may possess.

8·0 Temporary Works and Services, Attendance, Related Works

Temporary Accommodation

8·1 The Subcontractor shall provide to the approval of the Contractor and at his own expense, any requisite temporary site office, workshop accommodation, together with the necessary equipment, lighting, power, fuel etc.

Welfare facilities

8·2 The Subcontractor shall, at his own risk have reasonable and free use of the temporary welfare accommodation and/or services (including First Aid facilities and treatment) which the Contractor or the Employer may provide on the Site in connection with the Works.

Temporary services

8·3 The Subcontractor shall, at his own risk, have reasonable and free use, in common with others engaged upon the Site, of the water supply, temporary plumbing, temporary lighting and temporary electric power. Electric power supply for small tools and equipment used on the Site shall not exceed 110V A.C. single phase. Any electrical equipment used to carry out the Works must be in good mechanical condition and suitable for the electric power supply and fittings made available and fitted with suitable plugs, sockets and connectors to BS4343 (CEE 17) or any other standard that the Contractor may direct.

Use of Scaffolding

8·4 The Subcontractor shall at his own risk and at such time(s) and for such period(s) as the Contractor may direct have free use of the Contractor's scaffolding, ladders and mechanical hoisting facilities which may be available on the Site or already in position.

Delivery and Storage of Materials

8·5 The Subcontractor shall provide all materials, package and carriage to and from the Site. He will be responsible for unloading during the progress of his Works, storing in the areas provided and moving his own materials at the Site. Any materials delivered prior to commencement on Site shall be off-loaded by the Contractor at the sole risk and cost of the Subcontractor.

Removal of Rubbish etc.

8·6 All rubbish and/or surplus materials and plant of the Subcontractor must be removed forthwith from the vicinity of the Works, paths, roads etc., to an approved position on the Site.

Cutting Away

8·7 In no circumstances whatsoever shall any cutting away be done without the prior written authority of the Contractor.

Sub-surfaces

8·8 The Subcontractor shall satisfy himself before commencing work, as to the suitability of any surfaces to which the Subcontractor is to fix, apply or lay his work.

This Form is issued by the
Joint Council for Landscape Industries
comprising:-

Landscape Institute
Horticultural Trades Association
British Association Landscape Industries
National Farmers Unions
Institute of Leisure and Amenity Management.

Published for the Joint Council for Landscape Industries by the Landscape Institute
12 Carlton House Terrace, London SW1Y 5AH and available from RIBA Publications Limited
66 Portland Place, London W1N 4AD.

Printed by Duolith Ltd.
Welwyn Garden City, Herts.

Issued January 1986

Appendix C JLCI Standard Form of Agreement for Landscape Works

This Agreement is made the ________ day of ________ 19__

between ____________________

of ____________________
(hereinafter called 'the Employer')

of the one part AND ____________________

of (or whose registered office is at) ____________________

____

(hereinafter called 'the Contractor') of the other part.

Whereas

Recitals

1st the Employer wishes the following work ____________________

____

____

(hereinafter called "the Works") to be carried out under the direction of

____

(hereinafter called "Landscape Architect") and has caused drawings numbered
(hereinafter called "the Contract Drawings") [a] and/or a Specification (hereinafter called "the Contract Specification") [a] and/or schedules [a] and/or schedules of rates [a] and/or Bills of Quantities [a] which documents are together with the Conditions annexed hereto (hereinafter called "the Contract Documents")[a.1] showing and describing the work to be prepared and which are attached to this Agreement.

The law of England/Scotland shall be the proper law of this Agreement [a].

2nd The Contractor has stated the sum he will require for carrying out such work, which sum is that stated in Article 2 and has priced the Specification [a] or the schedules [a] or Bills of Quantities [a] or provided a schedule of rates [a];

3rd the Contract Documents have been signed by or on behalf of the parties hereto;

4th [a] the quantity surveyor appointed in connection with this contract shall mean

____

or in the event of his death or ceasing to be the quantity surveyor for this purpose such other person as the Employer nominates for that purpose;

---

[a] Delete as appropriate.
[a.1] Where a Contract Document has been priced by the Contractor it is this document that should be attached to this Agreement.

Now it is hereby agreed as follows

Article 1
For the consideration hereinafter mentioned the Contracter will in accordance with the Contract Documents carry out and complete the work referred to in the 1st Recital together with any changes made to that work in accordance with this Contract (hereinafter called "the Works").

Article 2
The Employer will pay the Contractor for the Works the sum of ____________________

____

______________________________________ (£ ______________________)

exclusive of VAT or such other sum as shall become payable hereunder at the times and in the manner specified in the Contract Documents.

Article 3
The term "Landscape Architect" in the said Conditions shall mean

____

or in the event of his death or ceasing to be the Landscape Architect for the purpose of this Contract such other person as the Employer shall within 14 days of the death or cessation as aforesaid nominate for that purpose, provided that no person subsequently appointed to be the Landscape Architect under this Contract shall be entitled to disregard or overrule any certificate or instruction given by the Landscape Architect for the time being.

Article 4
If any dispute or difference as to the construction of this agreement or any matter or thing of whatsoever nature arising thereunder or in connection therewith, except under the Supplementary Memorandum Part B clause B6 *(Value Added Tax)* or Part C *(Statutory tax deduction scheme)* to the extent provided in clause C8, shall arise between the Employer or the Landscape Architect on his behalf and the Contractor either during the progress or after the completion or abandonment of the Works or after the determination of the employment of the Contractor it shall be and is hereby referred to arbitration in accordance with clause 9. If under clause 9·1 the Employer and the Contractor have not agreed a person as the Arbitrator the appointor of the Arbitrator shall be the President or a Vice-President for the time being of the Landscape Institute.

As witness the hands of the parties hereto

Signed for and on behalf of the Employer ______________________________
(For the Employer)

in the presence of ______________________________
(Witness)

Signed for and on behalf of the Contractor ______________________________
(For the Contractor)

in the presence of ______________________________
(Witness)

Contents

Note:
*Clauses marked * should be completed/deleted as appropriate*
†see note on page 9

Conditions hereinbefore referred to

1·0 Intentions of the parties

Contractor's obligation

1·1 The Contractor shall with due diligence and in a good and workmanlike manner carry out and complete the Works in accordance with the Contract Documents using materials and workmanship of the quality and standards therein specified provided that where and to the extent that approval of the quality of materials or of the standards of workmanship is a matter for the opinion of the Landscape Architect such quality and standards shall be to the reasonable satisfaction of the Landscape Architect.

Landscape Architect's duties

1·2 The Landscape Architect shall issue any further information necessary for the proper carrying out of the Works, issue all certificates and confirm all instructions in writing in accordance with these Conditions.

Contract Bills and SMM

1·3 Where the Contract Documents include Contract Bills, the Contract Bills unless otherwise expressly stated therein in respect of any specified item or items are to have been prepared in accordance with the Standard Method of Measurement of Building Works, 7th Edition, published by the Royal Institution of Chartered Surveyors and the Building Employers Confederation (formerly National Federation of Building Trades Employers).

2·0 Commencement and completion

Commencement and completion

2·1 The Works may be commenced on

....

and shall be completed by

....

Extensions of contract period

2·2 If it becomes apparent that the works will not be completed by the date for completion inserted in clause 2·1 hereof (or any later date fixed in accordance with the provisions of this clause 2·2) for reasons beyond the control of the Contractor including compliance with any instruction of the Landscape Architect under this contract whose issue is not due to a default of the Contractor, then the Contractor shall thereupon in writing so notify the Landscape Architect who shall make, in writing, such extension of the time for completion as may be reasonable. Reasons within the control of the Contractor include any default of the Contractor or of others employed or engaged by or under him for or in connection with the works and of any supplier of goods or materials for the Works.

Damages for non-completion

2·3 If the Works are not completed by the completion date inserted in clause 2·1 hereof or by any later completion date fixed under clause 2·2 hereof then the Contractor shall pay or allow to the Employer liquidated damages at the rate of £............ per week for every week or part of a week between the aforesaid completion date and the date of practical completion. The Employer may deduct such liquidated damages from any monies due to the Contractor under this contract or he may recover them from the Contractor as a debt.

Practical Completion

2.4 The Landscape Architect shall certify the date when in his opinion the Works have reached practical completion.

Defects liability

2·5 Any defects, excessive shrinkages or other faults to the Works, other than tree, shrub, grass and other plant failures, which appear within three months [**b**].............. of the date of practical completion and are due to materials or workmanship not in accordance with the Contract or frost occuring before practical completion shall be made good by the Contractor entirely at his own cost unless the Landscape Architect shall otherwise instruct.
The Landscape Architect shall certify the date when in his opinion the Contractor's obligations under this clause 2·5 have been discharged.

Partial Possession by Employer

†2·6 If before practical completion of the Works the Employer with the consent of the Contractor shall take possession of any part then:
The date for practical completion of the part shall be the date of practical completion of the part and clause 4·3 shall apply to the part.
In lieu of the sum to be paid by the Contractor under clause 2·3 for any period during which the Works remain uncompleted after the Employers possession of any part the sum paid shall bear the same ration to the sum stated in clause 2·3 as does the Contract Sum less the value of the part to the Contract Sum.

Failures of Plants (Pre-Practical Completion)

†2·7 Any trees, shrubs, grass or other plants, other than those found to be missing or defective as a result of theft or malicious damage and which shall be replaced as set out in clause 6·5 A or B of these conditions, which are found to be or have been defective at practical completion of the works shall be replaced by the Contractor entirely at his own cost unless the Landscape Architect shall otherwise instruct. The Landscape Architect shall certify the dates when in his opinion the Contractor's obligations under this clause have been discharged.

(Post-Practical Completion)

*A The maintenance of trees, shrubs, grass and other plants after the date of Practical Completion will be carried out by the Contractor for the duration of the periods stated in accordance with the programme and in the manner specified in the Contract Documents. Any grass which is found to be defective within .... months, any shrubs, ordinary nursery stock trees or other plants found to be defective within . . . . months and any trees, semi-mature advanced or extra large nursery stock found to be defective within . . . . months of the date of Practical Completion and are due to materials or workmanship not in accordance with the contract shall be replaced by the Contractor entirely at his own cost unless the Landscape Architect shall otherwise instruct. The Landscape Architect shall certify the dates when in his opinion the Contractor's obligations under this clause have been discharged.

*B The maintenance of the trees, shrubs, grass and other plants after the date of the said certificate will be undertaken by the Employer who will be responsible for the replacement of any trees, shrubs, grass or other plants which are subsequently defective.

[**b**] If a different period is required delete 'three months' and insert the appropriate period.

* Delete (A) or (B) as appropriate.

3·0 Control of the Works

Assignment

3·1 Neither the Employer nor the Contractor shall, without the written consent of the other, assign this Contract.

Sub-contracting

3·2 The Contractor shall not sub-contract the Works or any part thereof without the written consent of the Landscape Architect whose consent shall not unreasonably be withheld.

Contractor's representative

3·3 The Contractor shall at all reasonable times keep upon the Works a competent person in charge and any instructions given to him by the Landscape Architect shall be deemed to have been issued to the Contractor.

Exclusion from the Works

3·4 The Landscape Architect may (but not unreasonably or vexatiously) issue instructions requiring the exclusion from the Works of any person employed thereon.

Landscape Architect's instructions

3·5 The Landscape Architect may issue written instructions which the Contractor shall forthwith carry out. If instructions are given orally they shall, in two days, be confirmed in writing by the Landscape Architect.
If within 7 days after receipt of a written notice from the Landscape Architect requiring compliance with an instruction the Contractor does not comply therewith then the Employer may employ and pay other persons to carry out the work and all costs incurred thereby may be deducted by him from any monies due or to become due to the Contractor under this Contract or shall be recoverable from the Contractor by the Employer as a debt.

Variations

†3·6 The Landscape Architect may, without invalidating the contract, order an addition to or omission from or other change in the Works or the order or period in which they are to be carried out and any such instruction shall be valued by the Landscape Architect on a fair and reasonable basis, using where relevant, prices in the priced specification/schedules/Bills of Quantity/schedule of rates [c] and such valuation shall include any direct loss and/or expense incurred by the Contractor due to the regular progress of the Works being affected by compliance with such instruction.
If any omission substantially varies the scope of the work such valuations shall take due account of the effect on any remaining items of work.

† Instead of the valuation referred to above, the price may be agreed between the Landscape Architect and the Contractor prior to the Contractor carrying out any such instruction.

P.C. and Provisional sums

3·7 The Landscape Architect shall issue instructions as to the expenditure of any P.C. and provisional sums and such instructions shall be valued or the price agreed in accordance with clause 3·6 hereof.

Objections to a Nomination

†3·8 The Landscape Architect shall not nominate any person as a nominated sub-contractor against whom the Contractor shall make reasonable objection or who will not enter into a sub-contract that applies the appropriate provisions of these conditions.

4.0 Payment

Correction of inconsistencies

4·1 Any inconsistency in or between the Contract Drawings [c] and the Contract Specification [c] and the schedules [c] shall be corrected and any such correction which results in an addition, omission or other change shall be treated as a variation under clause 3·6 hereof. Nothing contained in the Contract Drawings [c] or the Contract Specification [c] or the schedules or the Bills of Quantity [c] or the Schedule of Rates [c] shall override, modify or affect in any way whatsoever the application or interpretation of that which is contained in these Conditions.

Progress payments and retention

4·2* The Landscape Architect shall if requested by the Contractor, at intervals of not less than four weeks calculated from the date for commencement subject to any agreement between the parties as to stage payments, certify progress payments to the Contractor, in respect of the value of the Works properly executed, including any amounts either ascertained or agreed under clauses 3·6 and 3·7 hereof, and the value of any materials and goods which have been resonably and properly brought upon the site for the purpose of the Works and which are adequately stored and protected against the weather and other casualties less a retention of 5% (. . . . . %) [d] and less any sums previously certified, and the Employer shall pay to the Contractor the amount so certified within 14 days of the date of issue of the certificate.

Penultimate certificate

†4·3 The Landscape Architect shall within 14 days after the date of practical completion certified under clause 2·4 hereof certify payment to the Contractor of 97½% of the total amount to be paid to the Contractor under this contract so far as that amount is ascertainable at the date of practical completion, including any amounts either ascertained or agreed under clauses 3·6 and 3·7 hereof less the amount of any progress payments previously certified to the Employer, and less the cost of any subsequent mainten-ance included in the contract in accordance with Clause 2·7A unless the failures of plants as set out in clause 2·7 are in excess of 10% in which case the amount retained shall be adjusted accordingly, and the Employer shall pay to the Contractor the amount so certified within 14 days of the date of issue of that certificate.

Final certificate

4·4 The Contractor shall supply within three months/ . . . . [e] from the date of practical completion all documentation reasonably required for the computation of the amount to be finally certified by the Landscape Architect and the Landscape Architect shall within 28 days of receipt of such documentation, provided that the Landscape Architect has issued the certificate under clause 2·5 and 2·7A hereof, issue a final certificate certifying the amount remaining due to the Contractor or due to the Employer as the case may be and such sum shall as from the fourteenth day after the date of issue of the final certificate be a debt payable as the case may be by the Employer to the Contrac-tor or by the Contractor to the Employer.

---

[c] Delete as appropriate to follow any deletions in the recitals on page 1.

[d] If a different percentage is required delete 5% and insert the appropriate retention %.

[e] If a different period is required delete 'three months' and insert the appropriate period.

Contribution, levy and tax changes **[e1]**

4·5 Contribution, levy and tax changes shall be dealt with by the application of Part A of the Supplementary Memorandum to the Agreement for Landscape Works. The percentage addition under part A, clause A5 is .... % **[f]**.

Fixed price *

4·6A No account shall be taken in any payment to the Contractor under this Contract of any change in the cost to the Contractor of the labour, materials, plant and other resources employed in carrying out the Works except as provided in clause 4·5 hereof, if applicable.

Fluctuations *

4·6B The Contract sum shall be adjusted in accordance with the provisions of Part D of the Supplementary Memorandum and the Formula Rules current at the date stated in the tender documents, and which shall be incorporated in all certificates except those relating to the release of retention and shall be exclusive of any Value Added Tax. These provisions shall also be incorporated as appropriate in any sub contract agreement.

5·0 Statutory obligations

Statutory obligations, notices, fees and charges

5·1 The Contractor shall comply with, and give all notices required by, any statute, any statutory instrument, rule or order or any regulation or byelaw applicable to the Works (hereinafter called 'the statutory requirements') and shall pay all fees and charges in respect of the Works legally recoverable from him. If the Contractor finds any divergence between the statutory requirements and the contract documents or between the statutory requirements and any instruction of the Landscape Architect he shall immediately give to the Landscape Architect, a written notice specifying the divergence. Subject to this latter obligation, the Contractor shall not be liable to the Employer under this Contract if the Works do not comply with the statutory requirements where and to the extent that such non-compliance of the Work results from the Contractor having carried out work in accordance with the Contract Documents or any instruction of the Landscape Architect.

Value Added Tax

5·2 The sum or sums due to the Contractor under Article 2 hereof this Agreement shall be exclusive of any Value Added Tax and the Employer shall pay to the Contractor any Value Added Tax properly chargeable by the Commissioners of Custom and Excise on the supply to the Employer of any goods and services by the Contractor under this Contract in the manner set out in Part B of the Supplementary Memorandum to the Agreement for Landscape Works. Clause B1·1 of the Supplementary Memorandum, Part B applies/does not apply **[g]**

Statutory tax deduction scheme

5·3 Where at the date of tender the Employer was a 'Contractor', or where at any time up to the issue and payment of the final certificate the Employer becomes a 'Contractor', for the purposes of the statutory tax deduction scheme referred to in Part C of the Supplementary Memorandum to the Agreement for Landscape Works, Part C of that Memorandum shall be operated.

5·4 Number not used

Prevention of corruption

5·5 If the Employer is a local authority he shall be entitled to cancel this contract and to recover from the Contractor the amount of any loss resulting from such cancellation, if the Contractor shall have offered or given or agreed to give any person any gift or consideration of any kind or if the Contractor shall have committed any offence under the Prevention of Corruption Acts 1889 to 1916 or shall have given any fee or reward the receipt of which is an offence under sub-section (2) of section 117 of the Local Government Act 1972 or any re-enactment thereof.

6·0 Injury, damage and insurance

Injury to or death of persons

6·1 The Contractor shall be liable for and shall in demnify the Employer against any expense, liability, loss, claim or proceedings whatsoever arising under any statute or at common law in respect of personal injury to or death of any person whomsoever arising out of or in the course of or caused by the carrying out of the Works, except to the extent that the same is due to any act or neglect of the Employer or of any person for whom the Employer is responsible. Without prejudice to his liability to indemnify the Employer the Contractor shall take out and main tain and shall cause any sub-contractor to take out and maintain insurance which, in respect of liability to employees or apprentices shall comply with the Employer's Liability (Compulsory Insurance) Act 1969 and any statutory orders made thereunder or any amendment or re-enactment thereof and in respect of any other liability for personal injury or death shall be such as is necessary to cover the liability of the Con tractor, or, as the case may be, of such sub-contractor.

Injury or damage to property

6·2 The Contractor shall be liable for, and shall indemnify the Employer against, any expense, liability, loss, claim or proceedings in respect of any injury or damage whatsoever to any property real or personal (other than injury or damage to the Works) ot to any unfixed materials and goods delivered to, placed on or adjacent to the Works and intended therefor or, where clause 6·3B is applicable, to any property insured pursuant to clause 6·3B for the perils therein listed insofar as such injury or damage arises out of or in the course of or by reason of the carrying out of the Works and to the extent that the same is due to any negligence, breach of statutory duty, omission or default of the Contractor, his servants or agents, or of any person employed or engaged by the contractor upon or in connection with the Works or any part thereof, his servants or agents. Without prejudice to his obligation to indemnify the Employer the Contractor shall take out and maintain and shall cause any sub-contractor to

---

[e1] Delete clause 4·5 if the contract period is of such limited duration as to make the provisions of part A of the Supplementary Memorandum to this agreement inapplicable.

[f] Percentage to be inserted.

[g] Delete as required. Clause B1·1 can only apply where the Contractor is satisfied at the date the Contract is entered into that his output tax on **all** supplies to the Employer under the Contract will be at either a positive or a zero rate of tax.
On and from 1 April 1989 the supply in respect of a building designed for a 'relevant residential purpose' or for a 'relevant charitable purpose' (as defined in the legislation which gives statutory effect to the VAT changes operative from 1 April 1989) is only zero rated if the person to whom the supply is made has given to the Contractor a certificate in statutory form: see the VAT leaflet 708 revised 1989. Where a contract supply is zero rated by certificate only the person holding the certificate (usually the Contractor) may zero rate his supply.

take out and maintain insurance in respect of the liability referred to above in respect of injury or damage to any property real or personal other than the works which shall be for an amount not less than the sum stated below for any one occurence or series of occurences arising out of one event:

insurance cover referred to above to be not less than: £

Insurance of the Works by Contractor – Fire etc.

6·3A The Contractor shall in the joint names of Employer and Contractor insure the Works and all unfixed materials and goods delivered to, placed on or adjacent to the Works and intended therefor against loss and damage by fire, lightning, explosion, storm, tempest, flood, bursting or overflowing of water tanks, apparatus or pipes, earthquake, aircraft and other aerial devices or articles dropped therefrom, riot and civil commotion, for the full reinstatement value of the Works thereof plus . . . . . % [f] to cover professional fees.

Insurance of the Works and any existing structures by Employer – Fire etc.

6·3B The Employer shall in the joint names of Employer and Contractor insure against any loss or damage to any existing structures (together with the contents owned by him or for which he is responsible) and to the Works and all unfixed materials and goods intended for, delivered to, placed on or adjacent to the Works and intended therefore by fire, lightning, explosion, storm, tempest, flood, bursting or overflowing of water tanks, apparatus or pipes, earthquake, aircraft and other aerial devices or articles dropped therefrom, riot and civil commotion.

If any loss or damage as referred to in this clause occurs to the Works or to any unfixed materials and goods delivered to, placed on or adjacent to the Works and intended therefor then the Landscape Architect shall issue instructions for the reinstatement and making good of such loss or damage in accordance with clause 3·5 hereof and such instructions shall be valued under clause 3·6 hereof.

Evidence of insurance

6·4 The Contractor shall produce, and shall cause any sub-contractor to produce, such evidence as the Employer may reasonably require that the insurances referred to in clauses 6·1 and 6·2 and, where applicable 6·3A, hereof have been taken out and are in force at all material times. Where clause 6·3B hereof is applicable the Employer shall produce such evidence as the Contractor may reasonably require that the insurance referred to therein has been taken out and is in force at all material times.

Malicious Damage or Theft (before Practical Completion)

†6·5A All loss or damage arising from any theft or malicious damage prior to practical completion shall be made good by the Contractor at his own expense.

B The Contract sum shall include the provisional sum of £. . . . . [i] to be expended as instructed by the Landscape Architect in respect of the cost of all work arising from any theft or malicious damage to the works beyond the control of the Contractor prior to practical completion of the works.

7·0 Determination

Notices

7·1 Any notice or further notice to which clauses 7·2·1, 7·2·2, 7·3·1 and 7·3·2 refer shall be in writing and given by actual delivery, or by registered post or by recorded delivery. If sent by registered post or recorded delivery the notice or further notice shall, subject to proof to the contrary, be deemed to have been received 48 hours after the date of posting (excluding Saturday and Sunday and public holidays).

Determination by Employer

7·2·1 If the Contractor without reasonable cause makes default by failing to proceed diligently with the Works or by wholly or substantially suspending the carrying out the Works before practical completion the Landscape Architect may give notice to the Contractor which specifies the default and requires it to be ended. If the default is not ended within 7 days of receipt of the notice the Employer may by further notice to the Contractor determine the employment of the Contractor under this contract. Such determination shall take effect on the date of receipt of the further notice. A notice of determination under clause 7·2·1 shall not be given unreasonably or vexatiously.

7·2·2 If the Contractor
makes a composition or arrangement with his creditors, or becomes bankrupt, or
being a company,
makes a proposal for a voluntary arrangement for a composition of debts or scheme of arrangement to be approved in accordance with the Companies Act 1985 or the Insolvency Act 1986 as the case may be or any amendment or re-enactment thereof, or
has a provisional liquidator appointed, or
has a winding-up order made, or
passes a resolution for voluntary winding-up (except for the purposes of amalgamation or reconstruction), or
under the Insolvency Act 1986 or any amendment or re-enactment thereof has an administrator or an administrative receiver appointed
the Employer may by notice to the Contractor determine the employment of the Contractor under this Contract. Such determination shall take effect on the date of receipt of such notice.

7·2·3 Upon determination of the employment under clause 7·2·1 or clause 7·2·2 the Contractor shall immediately cease to occupy the site of the Works and the Employer shall not be bound to make any further payment to the Contractor that may be due under this Contract until after completion of the Works and the making good of any defects therein. The Employer may recover from the Contractor the additional cost to him of completing the Works, any expenses properly incurred by the Employer as a result of, and any direct loss and/or damage caused to the Employer by, the determination.

7·2·4 The provisions of clauses 7·2·1 to 7·2·3 are without prejudice to any other rights and remedies which the Employer may possess.

[h] Delete either clause 6·3A or clause 6·3B whichever clause is not required. If the Works:
– are not an extension to or an alteration to an existing structure, or
are an extension to or an alteration of an existing dwelling but the Employer cannot obtain the Insurance in accordance with clause 6·3B, clause 6·3B should be deleted.

[i] The amount to be inserted should take account of the Works and the place where they are carried out. Delete [A] or [B] as appropriate.

Determination by Contractor

7·3·1 If the Employer makes default in any one or more of the following respects:

1 he does not discharge in accordance with this Contract the amount properly due to the Contractor in respect of any certificate or pay any VAT due on that amount pursuant to clause 5·2 and the Supplementary Memorandum Part B;

2 he, or any person for whom he is responsible, interferes with or obstructs the issue of any certificate due under this Contract or interferes with or obstructs the carrying out of the Works or fails to make the premises available for the Contractor in accordance with clause 2·1 hereof;

3 he suspends the carrying out of the whole or substantially the whole of the Works for a continuous period of one month or more,

the Contractor may give notice to the employer which specifies the default and requires it to be ended. If the default is not ended within 7 days of receipt of the notice the Contractor may by further notice to the Employer determine the employment of the Contractor under this Contract. Such determination shall take effect on the date of receipt of the further notice. A notice of determination under clause 7·3·1 shall not be given unreasonably or vexatiously.

7·3·2 If the Employer
makes a composition or arrangement with his creditors, or becomes bankrupt, or
being a company,
makes a proposal for a voluntary arrangement for a composition of debts or scheme of arrangement to be approved in accordance with the Companies Act 1985 or the Insolvency Act 1986 as the case may be or any amendment or re-enactment thereof, or
has a provisional liquidator appointed, or
has a winding-up order made, or
passes a resolution for voluntary winding-up (except for the purposes of amalgamation or reconstruction), or
under the Insolvency Act 1986 or any amendment or re-enactment thereof has an administrator or an administrative receiver appointed the Contractor may by notice to the Employer determine the employment of the Contractor under this Contract. Such determination shall take effect on the date of receipt of such notice.

7·3·3 Upon determination of the employment of the Contractor under clause 7·3·1 or clause 7·3·2 the Contractor shall prepare an account setting out

- the total value of work properly executed and of materials and goods properly brought on the site for the purpose of the Works, such value to be ascertained in accordance with this Contract as if the employment of the Contractor had not been determined together with any amounts due to the Contractor under the Conditions not included in such total value; and
- the cost to the Contractor of removing or having removed from the site all temporary buildings, plant, tools and equipment; and
- any direct loss and/or damage caused to the Contractor by the determination.

After taking into account amounts previously paid or otherwise discharged to the Contractor under this Contract, the Employer shall pay to the Contractor the full amount properly due in respect of this account within 28 days of its submission by the Contractor.

7·3·4 The provisions of clauses 7·3·1 to 7·3·3 are without prejudice to any other rights and remedies which the Contractor may possess.

8·0 Supplementary Memorandum

Meaning of references in 4·5, 4·6, 5·2 and 5·3

8·1 The references in clauses 4·5, 4·6, 5·2 and 5·3 to the Supplementary Memorandum to the Agreement for Landscape Works are to that issued for use with this Form by the Joint Council for Landscape Industries as endorsed hereon.

9·0 Settlement of disputes – Arbitration

9·1 When the Employer or the Contractor require a dispute or difference as referred to in Article 4 to be referred to arbitration then either the Employer or the Contractor shall give written notice to the other to such effect and such dispute or difference shall be referred to the arbitration and final decision of a person to be agreed between the parties as the Arbitrator, or, upon failure so to agree within 14 days after the date of the aforesaid written notice, of a person to be appointed as the Arbitrator on the request of either the Employer or the Contractor by the person named in Article 4.

9·2 Subject to the provisions of clause A4·3 in the Supplementary Memorandum, the Arbitrator shall, whithout prejudice to the generality of his powers, have power to rectify the Agreement so that it accurately reflects the true agreement made by the Employer and the Contractor, to direct such measurements and/or valuations as may in his opinion be desirable in order to determine the rights of the parties and to ascertain and award any sum which ought to have been the subject of or included in any certificate and to open up, review and revise any certificate, opinion, decision, requirement or notice and to determine all matters in dispute which shall be submitted to him in the same manner as if no such certificate, opinion, decision, requirement or notice had been given.

9·3 The award of such Arbitrator shall be final and binding on the parties.

9·4 If before making his final award the Arbitrator dies or otherwise ceases to act as the Arbitrator, the Employer and the Contractor shall forthwith appoint a further Arbitrator, or, upon failure so to appoint within 14 days of any such death of cessation, then either the Employer or the Contractor may request the person named in Article 4 to appoint such further Arbitrator. Provided that no such further Arbitrator shall be entitled to disregard any direction of the previous Arbitrator or to vary or revise any award of the previous Arbitrator except to the extent that the previous Arbitrator had power so to do under the JCT Arbitration Rules and/or with the agreement of the parties and/or by the operation of law.

9·5 [**k**] The arbitration shall be conducted in accordance with the 'JCT Arbitration Rules' current at the date of this agreement [**l**]. Provided that if any amendments to the Rules so current have been issued by the Joint Contracts Tribunal after the aforesaid date the Employer and the Contractor may, by a joint notice in writing to the Arbitrator, state that they wish the arbitration to be conducted in accordance with the JCT Arbitration Rules as so amended.

[**k**] Delete clause 9·5 if it is not to apply.

[**l**] The JCT Arbitration Rules contain stricter time limits than those prescribed by some arbitration rules or those frequently observed in practice. The parties should note that a failure by a party or the agent of a party to comply with the time limits incorporated in these Rules may have adverse consequences.

September 1994 revision

Dated ______________________________ 19___

Agreement for Landscape Works

(In Scotland when English Law is not to apply supplementary clauses in respect of Building Regulations and Arbitration may be required)

First issued April 1978, revised April 1981 reprinted with corrections April 1982 revised April 1985, reprinted with corrections September 1986. Clause related to insurance provisions, revised September 1987. Revised January 1989, 90, 1991, reprinted with corrections to 2·3 and E1·1 January 1992, September 1992, 1st, 5th Recital Articles 1, and 2. Clauses 2, 2·2 and 2·5 revised, footnote (a1.) added.

Between ______________________________

____

and ______________________________

____

JCLI

This Form is issued by the Joint Council for Landscape Industries comprising:-

Landscape Institute
Horticultural Trades Association
British Association Landscape Industries
National Farmers' Union
Institute of Leisure and Amenity Management.

†Note: All clauses other than those marked † are similar to those in the Joint Contracts Tribunal Agreement for Minor Building Works.

Published for the Joint Council for Landscape Industries by the Landscape Institute
6/7 Barnard Mews, London SW11 1QU.
Available from RIBA Publications
Finsbury Mission, 39 Moreland Street,
London EC1V 8BB

Printed by Codicote Press Ltd.
Hitchin, Hertfordshire

The Joint Council for Landscape Industries has issued a Supplementary Memorandum for use with this Form, where applicable, as referred to in the list of Contents on page 3 and in clause 8·1. The Supplementary Memorandum is attached.

The Council previously issued Practice Note No. 1 and Practice Note No. 2, but both are superseded by Practice Note No. 3 which is attached.

The complete document now comprises 3 items:
Form of Agreement for Landscape Works.
Supplementary Memorandum.
JCLI Practice Note No. 3.
The Supplementary Memorandum and Practice Note No. 3 may be torn off as convenient.

Agreement for Landscape Works: Supplementary Memorandum

This Memorandum is the Supplementary Memorandum referred to in the Joint Council for Landscape Industries "Agreement for Landscape Works", 1981 Edition, clauses 4·5, 4·6, 5·2 and 5·3.

Unless otherwise specifically stated, words and phrases in the Supplementary Memorandum have the same meaning as in the Agreement for Landscape Works.

PART A – CONTRIBUTION, LEVY AND TAX CHANGES

Deemed calculation of Contract Sum – rates of contribution etc.

A1 The sum referred to in Article 2 (in this clause called "the Contract Sum") shall be deemed to have been calculated in the manner set out below and shall be subject to adjustment in the events specified hereunder:

A1·1 The prices used or set out by the Contractor in the contract documents are based upon the types and rates of contribution, levy and tax payable by a person in his capacity as an employer and which at the date of the contract are payable by the Contractor. A type and rate so payable are in clause A1·2 referred to as a 'tender type' and a 'tender rate'.

Increases or decreases in rates of contribution etc. – payment or allowance

A1·2 If any of the tender rates other than a rate of levy payable by virtue of the Industrial Training Act 1964, is increased or decreased, or if a tender type ceases to be payable, or if a new type of contribution, levy or tax which is payable by a person in his capacity as an employer becomes payable after the date of tender, then in any such case the net amount of the difference between what the Contractor actually pays or will pay in respect of

·1 workpeople engaged upon or in connection with the Works either on or adjacent to the site of the Works, and

·2 workpeople directly employed by the Contractor who are engaged upon the production of materials or goods for use in or in connection with the Works and who operate neither on nor adjacent to the site of the Works and to the extent that they are so engaged

or because of his employment of such workpeople and what he would have paid had the alteration, cessation or new type of contribution, levy or tax not become effective, shall, as the case may be, be paid to or allowed by the Contractor.

Persons employed on site other than 'workpeople'

A1·3 There shall be added to the net amount paid to or allowed by the Contractor under clause A1·2 in respect of each person employed on the site by the Contractor for the Works and who is not within the definition of 'workpeople' in clause A4·6·3 the same amount as is payable or allowable in respect of a craftsman under clause A1·2 or such proportion of that amount as reflects the time (measured in whole working days) that each such person is so employed.

A1·4 For the purposes of clause A1·3

no period less than 2 whole working days in any week shall be taken into account and periods less than a whole working day shall not be aggregated to amount to a whole working day;

the phrase "the same amount as is payable or allowable in respect of a craftsman" shall refer to the amount in respect of a craftsman employed by the Contractor (or by any sub-contractor under a sub-contract to which clause A3 refers) under the rules or decisions or agreements of the National Joint Council for the Building Industry or other wage-fixing body and, where the aforesaid rules or decisions or agreements provide for more than one rate of wage emolument or other expense for a craftsman, shall refer to the amount in respect of a craftsman employed as aforesaid to whom the highest rate is applicable; and

the phrase "employed . . . by the Contractor" shall mean an employment to which the Income Tax (Employment) Regulations 1973 (the PAYE Regulations) under section 204 of the Income and Corporation Taxes Act, 1970, apply.

Refunds and premiums

A1·5 The prices used or set out by the Contractor in the contract documents are based upon the types and rates of refund of the contributions, levies and taxes payable by a person in his capacity as an employer and upon the types and rates of premium receivable by a person in his capacity as an employer being in each case types and rates which at the date of tender are receivable by the Contractor. Such a type and such a rate are in clause A·6 referred to as a 'tender type' and a 'tender rate'.

A1·6 If any of the tender rates is increased or decreased or if a tender type ceases to be payable or if a new type of refund of any contribution levy or tax payable by a person in his capacity as an employer becomes receivable or if a new type of premium receivable by a person in his capacity as an employer becomes receivable after the date of tender, then in any such case the net amount of the difference between what the Contractor actually receives or will receive in respect of workpeople as referred to in clauses A1·2·1 and A1·2·2 or because of his employment of such workpeople and what he would have received had the alteration, cessation or new type of refund or premium not become effective, shall, as the case may be, be allowed by or paid to the Contractor.

A1·7 The references in clauses A1·5 and A1·6 to premiums shall be construed as meaning all payments howsoever they are described which are made under or by virtue of an Act of Parliament to a person in his capacity as an employer and which affect the cost to an employer of having persons in his employment.

Contracted-out employment

A1·8 Where employer's contributions are payable by the Contractor in respect of workpeople as referred to in clauses A1·2·1 and A1·2·2 whose employment is contracted-out employment within the meaning of the Social Security Pensions Act 1975 the Contractor shall for the purpose of recovery or allowance under this clause be deemed to pay employer's contributions as if that employment were not contracted-out employment.

Meaning of contribution etc.

A1·9 The references in clause A1 to contribution, levies and taxes shall be construed as meaning all impositions payable by a person in his capacity as an employer howsoever they are described and whoever the recipient which are imposed under or by virtue of an Act of Parliament and which affect the cost to an employer of having persons in his employment.

Materials – duties and taxes

A2·1 The prices used or set out by the Contractor in the contract documents are based upon the types and rates of duty if any and tax if any (other than any value added tax which is treated, or is capable of being treated, as input tax (as referred to in the Finance Act 1972) by the Contractor) by whomsoever payable which at the date of tender are payable on the import, purchase, sale, appropriation, processing or use of the materials, goods, electricity and, where so specifically stated in the Contract Documents, fuels specified in the list attached thereto under or by virtue of any Act of Parliament. A type and a rate so payable are in clause A2·2 referred to as a 'tender type' and a 'tender rate'.

A2·2 If in relation to any materials or goods specified as aforesaid, or any electricity or fuels specified as aforesaid and consumed on site for the execution of the Works including temporary site installations for those Works, a tender rate is increased or decreased or a tender type ceases to be payable or a new type of duty or tax (other than value added tax which is treated, or is capable of being treated as input tax (as referred to in the Finance Act 1972) by the Contractor) becomes payable on the import, purchase, sale, appropriation, processing or use of those materials, goods, electricity or fuels, then in any such case the net amount of the difference between what the Contractor actually pays in respect of those materials, goods, electricity or fuels and what he would have paid in respect of them had the alteration, cessation or imposition not occurred, shall, as the case may be, be paid to or allowed by the Contractor. In clause A2 the expression 'a new type of duty or tax' includes an additional duty or tax and a duty or tax imposed in regard to specified materials, goods, electricity or fuels in respect of which no duty or tax whatever was previously payable (other than any value added tax which is treated, or is capable of being treated, as input tax (as referred to in the Finance Act 1972) by the Contractor).

Fluctuations – work sublet

A3·1 If the Contractor shall decide to sublet any portion of the Works he shall incorporate in the sub-contract provisions to the like effect as the provisions of
clauses A1, A4 and A5 including the percentage stated in clause 4·5 pursuant to clause A5
which are applicable for the purposes of this Contract.

A3·2 If the price payable under such a sub-contract as aforesaid is decreased below or increased above the price in such sub-contract by reason of the operation of the said incorporated provisions, then the net amount of such decrease or increase shall, as the case may be, be allowed by or paid to the Contractor under this Contract.

Provisions relating to clauses A1, A3 and A5

Written notice by Contractor

A4·1 The Contractor shall give a written notice to the Landscape Architect of the occurrence of any of the events referred to in such of the following provisions as are applicable for the purposes of this Contract:
·1 clause A1·2
·2 clause A1·6
·3 clause A2·2
·4 clause A3·2

Timing and effect of written notices

A4·2 Any notice required to be given by the preceding sub-clause shall be given within a reasonable time after the occurrence of the event to which the notice relates, and the giving of a written notice in that time shall be a condition precedent to any payment being made to the Contractor in respect of the event in question.

Agreement – Landscape Architect and Contractor

A4·3 The Landscape Architect and the Contractor may agree what shall be deemed for all the purposes of this Contract to be the net amount payable to or allowable by the Contractor in respect of the occurrence of any event such as is referred to in any of the provisions listed in clause A4·1.

Fluctuations added to or deducted from Contract Sum – provisions setting out conditions etc. to be fulfilled before such addition or deduction

A4·4 Any amount which from time to time becomes payable to or allowable by the Contractor by virtue of clause A1 or clause A3 shall, as the case may be, be added to or deducted from the Contract Sum:

Provided:

– evidence by Contractor –

·1 As soon as is reasonably practicable the Contractor shall provide such evidence as the Landscape Architect may reasonably require to enable the amount payable to or allowable by the Contractor by virtue of clause A1 or clause A3 to be ascertained: and in the case of amounts payable to or allowable by the sub-contractor under clause A4·1·3 (or clause A3 for amounts payable to or allowable by the sub-contractor under provisions in the sub-contract to the like effect as clauses A1·3 and A1·4) – employees other than workpeople – such evidence shall include a certificate signed by or on behalf of the Contractor each week certifying the validity of the evidence reasonably required to ascertain such amounts.

– actual payment by Contractor –

·2 No amount shall be included in or deducted from the amount which would otherwise be stated as due in progress payments by virtue of this clause unless on or before the date as at which the total value of work, materials and goods is ascertained for the purposes of any progress payment the Contractor shall have actually paid or received the sum which is payable by or to him in consequence of the event in respect of which the payment or allowance arises.

– no alteration to Contractor's profit –

·3 No addition to or subtraction from the Contract Sum made by virtue of this clause shall alter in any way the amount of profit of the Contractor included in that Sum.

– position where Contractor is in default over completion

·4·1 No amount shall be included in or deducted from the amount which would otherwise be stated as due in progress payments or in the final certificate in respect of amounts otherwise payable to or allowable by the Contractor by virtue of clause A1 or clause A3 if the event (as referred to in the provisions listed in clause A4·1) in respect of which the payment or allowance would be made occurs after the completion date fixed under clause 2.

·4.2 Clause A4·4·4·1 shall not operate unless:
·1 the printed text of clause 2 is unamended, and
·2 the Landscape Architect has, in respect of every written notification by the Contractor under clause 2 of the Agreement, fixed such Completion Date as he considered to be in accordance with that clause.

Work etc. to which clauses A1 and A3 are not applicable

A4·5 Clause A1 and clause A3 shall not apply in respect of:

·1 work for which the Contractor is allowed daywork rates in accordance with any such rates included in the Contract Documents;

·2 changes in the rate of value added tax charged on the supply of goods or services by the Contractor to the Employer under this Contract.

Definitions for use with clause A1

A4·6 In clause A1:

·1 the expression 'the date of the contract' means the date 10 days before the date when the Agreement is executed by the parties;

·2 the expressions 'materials' and 'goods' include timber used in formwork but do not include other consumable stores, plant and machinery except electricity and, where specifically so stated in the Contract Documents, fuels;

·3 the expression 'workpeople' means persons whose rates of wages and other emoluments (including holiday credits) are governed by the rules or decisions or agreements of the National Joint Council for the Building Industry or some other wage-fixing body for trades associated with the building industry;

·4 the expression 'wage-fixing body' means a body which lays down recognised terms and conditions of workers within the meaning of the Employment Protection Act 1975, Schedule II, paragraph 2(a).

Percentage addition to fluctuation payments or allowances

A5 There shall be added to the amount paid or allowed by the Contractor under
·1 clause A1·2
·2 clause A1·3
·3 clause A1·6
·4 clause A2·2
the percentage stated in clause 4·5.

PART B – VALUE ADDED TAX

B1 In this clause 'VAT' means the value added tax introduced by the Finance Act 1972 which is under the care and management of the Commissioners of Customs and Excise (hereinafter called the 'Commissioners').

B1·1 ·1 Where in clause 5·2 it is stated that clause B1·1 applies clauses B2·1 and B2·2 hereof shall not apply unless and until any notice issued under clause B1·1·4 hereof becomes effective or unless the Contractor fails to give the written notice required under clause B1·1·2. Where clause B1·1 applies clauses B1 and B3·1 to B10 inclusive remain in full force and effect.

·2 Not later than 7 days before the date for the issue of the first certificate under clause 4·2 the Contractor shall give written notice to the Employer, with a copy to the Landscape Architect, of the rate of tax chargeable on the supply of goods and services for which certificates under clauses 4·2 and 4·3 and the final certificate under clause 4·4 are to be issued. If the rate of tax so notified is varied under statute the Contractor shall not later than 7 days after the date when such varied rate comes into effect, send to the Employer, with a copy to the Landscape Architect, the necessary amendment to the rate given in his written notice and that notice shall then take effect as so amended.

·3 For the purpose of complying with clause 5·2 for the payment by the Employer to the Contractor of tax properly chargeable by the Commissioners of Customs and Excise on the Contractor, an amount calculated at the rate given in the aforesaid written notice (or, where relevant, amended written notice) shall be shown on each certificate issued by the Landscape Architect under clauses 4·2 and 4·3 and, unless the procedure set out in clause B3 hereof shall have been completed, on the final certificate issued by the Landscape Architect under clause 4·4. Such amount shall be paid by the Employer to the Contractor or by the Contractor to the Employer as the case may be written the period for payment of certificates given in clauses 4·2, 4·3 and 4·4.

·4 Either the Employer or the Contractor may give written notice to the other, with a copy to the Landscape Architect, stating that with effect from the date of the notice clause B1·1 shall no longer apply. From that date the provisions of clauses B2·1 and B2·2 shall apply in place of clause B1·1 hereof.

B2·1 Unless clause B1·1 applies the Landscape Architect shall inform the Contractor of the amount certified under clause 4·2 and immediately the Contractor shall give to the Employer a written provisional assessment of the respective values of those supplies of goods and services for which the certificate is being issued and which will be chargeable at the relevant times of supply on the Contractor at any rate or rates of VAT (including zero). The Contractor shall also specify the rate or rates of VAT which are chargeable on those supplies.

B2·2 Upon receipt of the Contractor's written provisional assessment the Employer shall calculate the amount of VAT due by applying the rate or rates of VAT specified by the Contractor to the amount of the supplies included in his assessment, and shall remit the calculated amount of such VAT to the Contractor when making payment to him of the amount certified by the Landscape Architect under clause 4·2.

B3·1 Where clause B1·1 is operated clause B3 only applies if no amount of tax pursuant to clause B1·1·3 has been shown on the final certificate issued by the Landscape Architect. After the issue by the Landscape Architect of his certificate of making good defects under clause 2·5 of the Agreement the Contractor shall, as soon as he can finally so ascertain, prepare and submit to the Employer a written final statement of the value of all supplies of goods and services for which certificates have been or will be issued which are chargeable on the Contractor at any rate or rates of VAT (including zero). The Contractor shall also specify the rate or rates of VAT which are chargeable on those supplies and shall state the grounds on which he considers such supplies are so chargeable. He shall also state the total amount of VAT already received by him.

B3·2 Upon receipt of the written final statement the Employer shall calculate the amont of VAT due by applying the rate or rates of VAT specified by the Contractor to the value of the supplies included in the statement and deducting therefrom the total amount of VAT already received by the Contractor and shall pay the balance of such VAT to the Contractor within 28 days from receipt of the statement.

B3·3 If the Employer finds that the total amount of VAT specified in the final statement as already paid by him exceeds the amount of VAT calculated under clause B3·2, he shall so notify the Contractor, who shall refund such excess to the Employer within 28 days of receipt of the notification together with a receipt under clause B4 hereof showing a correction of the amounts for which a receipt or receipts have previously been issued by the Contractor.

B4 Upon receipt of any VAT properly paid under the provisions of this clause the Contractor shall issue to the Employer an authenticated receipt of the kind referred to in Regulation 12(4) of the Value Added Tax (General) Regulations 1985 or any amendment or re-enactment thereof.

B5·1 In calculating the amount of VAT to be paid to the Contractor under clauses B2 and B3 hereof, the Employer shall disregard any sums which the Contractor may be liable to pay or allow to the Employer, or which the Employer may deduct, under clause 2·3 as liquidated damages, as liquidated damages under clause 2·3.

B5·2 The Contractor shall likewise disregard such liquidated damages when stating the value of supplies of goods or services in his written statement under clause B3·1.

B5·3 Where clause B1·1 is operated the Employer shall pay the tax to which that clause refers notwithstanding any deduction which the Employer may be empowered to make by clause 2·3 from monies due to the Contractor under certificates for payment issued by the Landscape Architect.

B6·1 If the Employer disagrees with te final statement issued by the Contractor under clause B3·1 he may request the Contractor to obtain the decision of the Commissioners on the VAT properly chargeable on the Contractor for all supplies of goods and services under this contract and the Contractors shall forthwith request the Commissioners for such decision.

B6·2 If the Employer disagrees with such decision, then, provided he secures the Contractor against all costs and other expenses, the Contractor shall in accordance with the instructions of the Employer make all such appeals against the decision of the Commissioners as the Employer may request.

B6·3 Within 28 days of the date of the decision of the Commissioners (or of the final adjudication of an appeal) the Employer or the Contractor as the case may be, shall pay or refund to the other any VAT underpaid or overpaid in accordance with such decisions or adjudication. The Contractor shall also account to the Employer for any costs awarded in his favour. The provisions of clause B3·1 shall apply in regard to the provision of authenticated receipts.

B7 The provisions of clause 8 shall not apply to any matters to be dealt with under clause B6.

B8 If any dispute or difference between the Employer and the Contractor is referred to an Arbitrator appointed under clause 9 or to a Court, then insofar as any payment awarded in such arbitration or court proceedings varies amounts certified for payment of goods or services supplied by the Contractor to the Employer under this Contract or is an amount which ought to have been but was not so certified, then the provisions of this Part B shall so far as relevant and applicable apply to any such payments.

B9 Notwithstanding any provisions to the contrary elsewhere in the Agreement the Employer shall not be obliged to make any further payment to the Contractor if the Contractor is in default in providing the receipt referred to in clause B4; provided that this clause B9 shall only apply where

the Employer can show that he requires such receipt to validate any claim for credit for tax paid or payable under this Part B which the Employer is entitled to make to the Commissioners and

the Employer has paid tax in accordance with the provisional assessment of the Contractor under clause B2·2 or paid tax in accordance with clause B1·1.

B10 The Employer shall be discharged from any further liability to pay tax to the Contractor under the clause upon payment of tax in accordance with with clause B3·2 (adjusted where relevant in accordance with the decision in any appeal to which clause B6 refers) or with clause B1·1·3 in respect of the tax shown in the final certificate. Provided always that if after the due discharge under clause B10 the Commissioners decide to correct the tax due from the Contractor on the supply to the Employer of any goods and services by the Contractor under this Contract the amount of such correction shall be an additional payment by the Employer to the Contractor or by the Contractor to the Employer as the case may be. The provisions of clause B6 in regard to disagreement with any decision of the Commissioners shall apply to any decision referred to in this proviso.

PART C – STATUTORY TAX DEDUCTION SCHEME – Income and Corporation Taxes Act, 1988

C1 In this clause 'the Act' means the Income and Corporation Taxes Act, 1988 and 'the Regulations' means the Income Tax (Sub-Contractors in the Construction Industry) Regulations 1985 S.I. No. 1960 or any re-enactment or amendment or remaking thereof; 'contractor' means a person who is a contractor for the purposes of the Act and the Regulations; 'evidence' means such evidence as is required by the Regulations to be produced to a 'contractor' for the verification of a 'sub-contractor's' tax certificate; 'statutory deduction' means the deduction referred to in section 559(4) of the Act or such other deduction as may be in force at the relevant time; 'sub-contractor' means a person who is a sub-contractor for the purposes of the Act and the Regulations; 'tax certificate' is a certificate issuable under the Act.

Provision of evidence – tax certificate

C2·1 Not later than 21 days before the first payment becomes due under clause 4 or after the Employer becomes a 'contractor' as referred to in clause 5·3 the Contractor shall:

either

·1 provide the Employer with the evidence that the Contractor is entitled to be paid without the statutory deduction;

or

·2 inform the Employer in writing, and send a duplicate copy to the Landscape Architect, that he is not entitled to be paid without the statutory deduction.

C2·2 If the Employer is not satisfied with the validity of the evidence submitted in accordance with clause C2·1·1 hereof, he shall within 14 days of the Contractor submitting such evidence notify the Contractor in writing that he intends to make the statutory deduction from payments due under this Contract to the Contractor who is 'a sub-contractor' and give his reasons for that decision. The Employer shall at the same time comply with clause C5·1.

Uncertificated Contractor obtains tax certificate

C3·1 Where clause C2·1·2 applies, the Contractor shall immediately inform the Employer if he obtains a tax certificate and thereupon clause C2·1·1 shall apply.

Expiry of tax certificate

C3·2 If the period for which the tax certificate has been issued to the Contractor expires before the final payment is made to the Contractor under this Contract the Contractor shall, not later than 28 days before the day of expiry:

either

·1 provide the Employer with evidence that the Contractor from the said date of expiry is entitled to be paid for a further period without the statutory deduction in which case the provisions of clause C2·2 hereof shall apply if the Employer is not satisfied with the evidence;

or

·2 inform the Employer in writing that he will not be entitled to be paid without the statutory deduction after the said date of expiry.

Cancellation of tax certificate

C3·3 The Contractor shall immediately inform the Employer in writing if his current tax certificate is cancelled and give the date of such cancellation.

---

[a] The application of the Tax Deduction scheme and these provisions is explained in JCT Practice Note No. 22.

Vouchers

C4 The Employer shall, as a 'contractor', in accordance with the Regulations, send promptly to the Inland Revenue any voucher which, in compliance with the Contractor's obligations as a 'sub-contractor' under the Regulations, the Contractor gives to the Employer.

Statutory deduction – direct cost of materials

C5·1 If at any time the Employer is of the opinion (whether because of the information given under clause C2·1·2 of this clause or of the expiry or cancellation of the Contractor's tax certificate or otherwise) that he will be required by the Act to make a statutory deduction from any payment due to be made the Employer shall immediately so notify the Contractor in writing and require the Contractor to state not later than 7 days before each future payment becomes due (or within 10 days of such notification if that is later) the amount to be included in such payment which represents the direct cost to the Contractor and any other person of materials used or to be used in carrying out the Works.

C5·2 Where the Contractor complies with clause C5·1 he shall indemnify the Employer against loss or expense caused to the Employer by any incorrect statement of the amount of direct cost referred to in that clause.

C5·3 Where the Contractor does not comply with clause C5·1 the Employer shall be entitled to make a fair estimate of the amount of direct cost referred to in that clause.

Correction of errors

C6 Where any error or omission has occured in calculating or making the statutory deduction the Employer shall correct that error or omission by repayment to, or by deduction from payments to, the Contractor as the case may be, subject only to any statutory obligation on the Employer not to make such correction.

Relation to other Clauses of Agreement

C7 If compliance with this clause involves the Employer or the Contractor in not complying with any other provisions of the Agreement, then the provisions of this clause shall prevail.

Application of Arbitration Agreement

C8 The provisions of Article 4 (arbitration) shall apply to any dispute or difference between the Employer or the Landscape Architect on his behalf and the Contractor as to the operation of this clause except where the Act or the Regulations or any other Act of Parliament or statutory instrument rule or order made under an Act of Parliament provide for some other method of resolving such dispute or difference.

PART D – FLUCTUATIONS

Procedure for the recovery from the Employer of any Fluctuations in the cost of labour and materials (*delete if not applicable*).

Adjustment of Contract Sum — NEDO Price Adjustment Formula for Landscape Contracts

D1·1 The Contract Sum shall be adjusted in accordance with the provisions of this clause and the Formula Rules current at the Date of Tender issued for use with this clause by the Joint Contracts Tribunal for the Standard Form of Building Contract hereinafter called 'the Formula Rules'.

D1·2 Any adjustment under this clause shall be to sums exclusive of value added tax and nothing in this clause shall affect in any way the operation of Clause 5·2 (value added tax) and Part B of this Memorandum.

D2 The Definitions in rule 3 of the Formula Rules shall apply to this clause.
The adjustment referred to in this clause shall be effected (after taking into account any Non-Adjustable Element) in all certificates for payment (other than those under Clause 4·2 of these Conditions (release of retention)) issued under the provisions of these Conditions.
If any correction of amounts of adjustment under this clause included in previous certiificates is required following any operation of rule 5 of the Formula Rules such correction shall be given effect in the next certificate for payment to be issued.

Fluctuations – Articles manufactured outside the United Kingdom

D3 For any article to which rule 4·2 of the Formula Rules applies the Contractor shall insert in a list attached to the Contract Documents the market price of the article in sterling (that is the price delivered to the site) current at the Date of Tender. If after that Date the market price of the article inserted in the aforesaid list increases or decreases then the net amount of the difference between the cost of purchasing at the market price inserted in the aforesaid list and the market price payable by the Contractor and current when the article is bought shall, as the case may be, be paid to or allowed by the Contractor. The reference to 'market price' in this clause shall be construed as including any duty or tax (other than any value added tax which is treated, or is capable of being treated, as input tax (as defined in the Finance Act 1972) by the Contractor) by whomsoever payable under or by virtue of any Act of Parliament on the import, purchase, sale, appropriation or use of the article specified as aforesaid.

Power to agree – Landscape Architect and Contractor

D4 The Landscape Architect and the Contractor may agree any alteration to the methods and procedures for ascertaining the amount of formula adjustment to be made under this clause and the amounts ascertained after the operation of such agreement shall be deemed for all the purposes of this Contract to be the amount of formula adjustment payable to or allowable by the Contractor in respect of the provisions of this Clause. Provided always:

D4·1 that no alteration to the methods and procedures shall be agreed as aforesaid unless it is reasonably expected that the amount of formula adjustment so ascertained will be the same or approximately the same as that ascertained in accordance with Part I or Part II of Sections 2 of the Formula Rules whichever Part is stated to be applicable in the Contract Documents, and

D4·2 that any agreement under this sub-clause shall not have any effect on the determination of any adjustment payable by the Contractor to any sub-contractor.

Position where Monthly Bulletins are delayed, etc.

D5·1 If at any time prior to the issue of the Final Certificate under Clause 4·3 of these Conditions formula adjustment is not possible because of delay in or cessation of, the publication of the Monthly Bulletins, adjustment of the Contract Sum shall be made in each Interim Certificate during such period of delay on a fair and reasonable basis.

D5·2 If publication of the Monthly Bulletins is recommenced at any time prior to the issue of the Final Certificate under Clause 4·3 of these Conditions the provisions of this clause and the Formula Rules shall operate for each Valuation Period as if no delay or cessation as aforesaid had occurred and the adjustment under this clause and the Formula Rules shall be substituted for any adjustment under paragraph D5·1 hereof.

D5·3 During any period of delay or cessation as aforesaid the Contractor and Employer shall operate such parts of this clause and the Formula Rules as will enable the amount of formula adjustment due to be readily calculated upon recommencement of publication of the Monthly Bulletins.

Formula Rules

Rule 3

D6·1 Base Month................................... 19.........
Non-Adjustable Element.............% (not to exceed 10%).

Rules 10 and 30(i)

*Part I/Part II of Section 2 of the Formula Rules is to apply.

---

* *Strike out according to which method of formula adjustment (Part I – Work Category Method or Part II – Work Group Method) has been notified in the Schedules of Quantity issued to tenderers.*

This Supplementary Memorandum is issued by the Joint Council for Landscape Industries comprising:-

Landscape Institute
Horticultural Trades Association
British Association Landscape Industries
National Farmers' Union
Institute of Leisure and Amenity Management

Published for the Joint Council for Landscape Industries by the Landscape Institute
6/7 Barnard Mews, London SW11 1QU.
Available from RIBA Publications
Finsbury Mission, 39 Moreland Street, London EC1V 8BB

Printed by Codicote Press Ltd., Hitchin, Hertfordshire

First issued April 1985, revised January 1991, 1992 and September 1994 (C1 revised)

Practice Note No. 3 (April 1985) JCLI Standard Form of Agreement for Landscape Works

September 1994 revision

Note: The JCLI Practice Note 3 supersedes No. 1 issued April 1978 and No. 2 on April 1982.

1. **Schedules (1st Recital)** means a list of items giving quantities, as necessary, and descriptions of work required, prepared in accordance with an appropriate method of measurement, which will be priced by tenderers and subsequently form part of the contract. Detailed information on materials and workmanship shall be contained in the Specification.

   Bills of Quantity
   Provision is now made for the option of including in the contract documents Bills prepared in accordance with SMM and the JCLI Rules.

   4th recital
   This provides for the naming of a quantity surveyor; neither the Articles nor the Conditions mention him but the Employer or the Landscape Architect on his behalf may wish to appoint a quantity surveyor in connection with the project, for example, to assist with valuations.

   Article 2
   The reference to VAT is new.

   Article 4: Arbitration
   Follows clause 8 (1978).

   Page 3
   The top part of the page is left blank for the parties to seal the Agreement if so required.

2. **Liquidated and Ascertained Damages: (Clause 2·3)**

   In landscape work it is generally very difficult to define the loss an Employer will suffer when a contract is delayed by the Contractor.

   The method outlined below for assessing the damages to be included in the contract is based on a notional interest on employed capital for the period of delay.

   The value of the project is assessed and interest at 1% over Bank Minimum Lending Rate is charged. This is divided by 52 to realise a rate per week.

   Example:
   Contract Sum £75,000
   M.L.R. 12½%
   Interest 13½%

   Liquidated and Ascertained Damages $\frac{£75{,}000 \times 13{\cdot}5}{52 \times 100} = £194.71$ per week

The essence of this method is that the damages are both ascertainable and realistic, being based on capital employed and not on some arbitrary assessment. Note: It does not however make any allowance for any additional professional fees which may be incurred as a result of the extended contract period.

3. **Date of Practical Completion: (Clause 2·4)**

   The date of Practical Completion is that date when the Landscape Architect certifies that the Contractor has fulfilled his obligations under the terms of the Contract. This does not preclude partial possession (dealt with under Clause 2·6) but simply determines the commencement of the Defects Liability Period for sections of work.

4. **Defects Liability: (Clause 2·5)**

   Except for plants which are dealt with separately the Contractor is responsible for making good all defects which appear after Practical Completion within a period inserted in the contract. It is recommended that the liability periods for landscape work should differ for hard landscape, grass, shrubs and trees. It is recommended that for hard landscape this period should normally be 6 months. Where paving forms a minor element within a predominantly soft landscape then the period should be as for the associated planting. In all cases the period should begin on the date of Practical Completion.

5. **Plant Failures – at Practical Completion: (Clause 2·7)**

   All plants which have failed prior to Practical Completion or within 6 weeks of the first leafing out in the case of trees and shrubs or grass at the first cut and roll, whichever is the later, are to be replaced by the Contractor entirely at his own expense, unless it arises from theft or malicious damage after Practical Completion. At an appropriate time the Contractor and Landscape Architect should meet to establish the extent of any defects and failures. The Landscape Architect will then prepare a schedule of these defects and send it to the Contractor within 14 days. The Contractor will inform the Landscape Architect when the work has been rectified.

6. **Plant Failures – after Practical Completion: (Clause 2·7(A) or (B))**

   At the time of preparing the tender documents a decision must be made as to whether the Contractor or the Employer is to be responsible for Maintenance of trees, shrubs and grass areas after Practical Completion. Maintenance is defined as those operations carried out by the Contractor or Employer after Practical Completion.

   i. If the Employer is to be responsible for Maintenance following Practical Completion then 2·7(A) is deleted and 2·7(B) is used, the Contractor is then relieved of all further obligations to replace defective plants other than those items listed in the Schedule of Defects referred to in item 5 above.

ii. If Maintenance is to be carried out by the Contractor 2·7(B) is deleted and 2·7(A) is used and the Contractor is then responsible for making good all defective plants during the Period, and this maintenance is to be separately scheduled and priced in the Schedules of Quantity. The Landscape Architect must not state for example, 'the Contractor is to allow for all maintenance during the Defects Liability Period', but must specify the operations in the Schedules of Quantity so that the Contractor is fully aware of his commitments and can incorporate the cost of such work in the Contract Sum. This does not include the cost of making good defects at Practical Com pletion which is deemed to be incorporated in the Contractor's Contract Sum. When the Contractor is responsible for Maintenance the following Defects Liability Periods for plants are recommended:

Grass: 3 months after the first cut and roll.
Shrubs and Trees: 12 months after the first breaking of buds.
Semi mature and ELNS: 2 years after the first breaking of buds.

In all cases the period should begin on the date of Practical Completion.

It is recommended that where varying Defects Liability Periods are required by the Contract, work relating to each Period shall be kept separate under a suitable heading, which can be identified with the particular periods, and only that part of the retention monies proportional to the value of the part of the work certified as being practically complete should be released.

7. P.C. Sums: (Clause 3·7)

Particular attention is drawn to Clause 3·6 where it will be noted that no cash discount is allowable by sub-contractors or suppliers arising out of instructions issued by the Landscape Architect in respect of prime cost sums. As such quotations are accepted nett Contractors are deemed to have included for the lack of cash discount elsewhere in the Contract Sum.

8. Correction of inconsistencies (Clause 4·1)

This does not provide that every correction is to be treated as a variation but only any correction of an inconsistency which results in a change. Where the correction is not a change there is no variation. It may be necessary to determine which of two inconsistent documents is the ruling document or which of two inconsistent statements prevails. For example, if written descriptions appear on a drawing and it is stated in the Contract Documents that any figure in a description shall prevail over any figures otherwise shown on a drawing there would only be a change if it were necessary to correct a figure in a description where the inconsistent figure on the drawing was actually required.

9. Malicious Damage: (Clause 6·5)

The cost of all making good arising from malicious damage or theft after Practical Completion is always the responsibility of the Employer. When the risk of such damage is considered small, if the Landscape Architect wishes such damage or theft prior to Practical Completion to be made good by the Contractor at no extra cost to the Contract as at present, Clause 6·5(A) should be used. The Contractor will then make his own assessment of the likely cost and include for it in his tender and which the Employer then pays whether any damage arises or not.

If it is preferred to reimburse the Contractor only for the actual cost of such damage which arises prior to Practical Completion then a provisional sum should be included in the tender documents and Clause 6·5(B) used. During the Defects Liability period, the Landscape Contractor is to advise the Landscape Architect each month of any plants found to be dead, in an unhealthy condition and those stolen or damaged by vandalism. The Landscape Architect will then issue the necessary instructions for their replacement at the Employers or Landscape Contractors expense as applicable.

10. Supplementary Memorandum (Clause 8·0)

Part A of the Supplement, Tax etc. changes, corresponds with clause 38 of the JCT Standard Form of Building Contract (1980); Part B, VAT, corresponds with the VAT Agreement Supplement to the Standard Form; and Part C, Statutory Tax Deduction Scheme, corresponds with clause 31 of the Standard Form.

11. Temporary Protection:

If temporary protective measures including fencing are required, these must be clearly stated and quantified at tender stage. It is not sufficient to state the 'the Contractor is to allow for the erection of any temporary fencing which may be necessary to protect grass or shrub areas.' Ownership of temporary protection may often with advantage be transferred to the Employer if so stated in the tender documents otherwise it remains the property of the Contractor.

12. Watering

The tender and contract documents should state the frequency and quantity of water required to be provided by the Contractor. In drought conditions when the provision of water is restricted by legislation contractors and sub-contractors should be required to ascertain from the water authority or any other source if second class water is available and if so should be required to inform the Landscape Architect of the cost and await instructions.

13. Maintenance

When Maintenance is the responsibility of the Contractor in accordance with Clause 2·7A the value of this work is excluded from that included in the penultimate certificate issued in accordance with clause 4·3.

This Practice Note is issued by the
Joint Council for Landscape Industries comprising:

Landscape Institute
Horticultural Trades Association
British Association Landscape Industries
National Farmers' Union
Institute of Leisure and Amenity Management

Published for the Joint Council for Landscape Industries
by the Landscape Institute
6/7 Barnard Mews, London SW11 1QU.
Available from RIBA Publications
Finsbury Mission, 39 Moreland Street,
London EC1V 8BB

Printed by Codicote Press Ltd
Hitchin, Hertfordshire

First issued April 1985, revised January 1991,
January and September 1992, September 1994

Appendix D JLCI Supplement to the JCT Intermediate Form IFC 84

Practice Note No. 4 (APRIL 1985) JCLI Supplement to the JCT Intermediate Form IFC 84

The JCLI Agreement for Landscape Works was first issued in April 1978 at which time the 1968 JCT Minor Works Form on which it was based had no provision for 2·7 Plant Failures nor 6·5 Malicious Damage, neither did it cover:-

- 2·6 Partial possession
- 3·6 The valuation of variations using tender rates
- 3·8 P.C. Sums and objections to nominations
- 3·9 Direct loss and expense if progress is disturbed
- 4·6 Fluctuations

detailed provisions were therefore incorporated to cover these aspects.

While suitable for the majority of landscape contracts over the six years since it was first published, instances have occurred of Local Authorities expressing reluctance to use the Form on landscape contracts in excess of £75,000.

The JCT has now published an "Intermediate" Form (IFC 84) for use on all contracts both private and public sector local authority, with or without quantities prepared in accordance with the Standard Method of Measurement (SMM6). While it still does not cover Plant Failures and Malicious Damage, the remaining aspects previously omitted are now provided for.

For those larger landscape contracts the detailed provision of the JCT Intermediate Form may be appropriate and the attached landscape supplementary clauses should be incorporated.

This introductory page may be torn off

JCLI Supplement to be read in conjunction with the JCT Intermediate Form (IFC84)

Clause 2·0 Possession and Completion

Add new clause 2·11

Partial Possession by the Employer

2·11 If any at any time before Practical Completion of the Works the Employer with the consent of the Contractor shall take possession of any part or parts of the same (any such part being referred to in this clause 2·11 as 'the relevant part') then notwithstanding anything expressed or implied elsewhere in this Contract:

- for the purpose of clause 2·10 *(Defects liability)* and 4·3 *(Interim payment)* Practical Completion of the relevant part shall be deemed to have occurred and the defects liability period in respect of the relevant part shall be deemed to have commenced on the date on which the Employer shall have taken possession thereof;
- as from the date which the Employer shall have taken possession of the relevant part, the obligation of the Contractor to insure under clause 6·3A, if applicable, shall terminate in respect of the relevant part but not further or otherwise;
- in lieu of any sum to be paid or allowed by the Contractor under clause 2·7 *(Liquidated damages)* in respect of any period during which the Employer shall have taken possession of the relevant part there shall be paid or allowed such sum as bears the same ratio to the sum which would be paid or allowed apart from the provisions of this clause 2·11 as the Contract Sum, less the amount contained therein in respect of the said relevant part, bears to the Contract Sum.

Add new clause 2·12

Failures of Plants (Pre-Practical Completion)

2·12 Any trees, shrubs, grass or other plants, other than those found to be missing or defective as a result of theft or malicious damage and which shall be replaced as set out in clause 6·4 of these conditions, which are found to be defective at practical completion of the works shall be replaced by the Contractor entirely at his own cost unless the Landscape Architect shall otherwise instruct. The Landscape Architect shall certify the dates when in his opinion the Contractor's obligation under this clause have been discharged.

(Post-Practical Completion)

*A The maintenance of trees, shrubs, grass and other plants after the date of the said certificate will be carried out by the Contractor for the duration of the periods stated in accordance with the programme and in the manner specified in the Contract Documents. Any grass which is found to be defective within.......... months, any shrubs, ordinary nursery stock trees or other plants found to be defective within .... months and any trees, semi-mature advanced or extra large nursery stock found to be defective within .... months of the date of Practical Completion and are due to materials or workmanship not in accordance with the contract shall be replaced by the Contractor entirely at his own cost unless the Landscape Architect shall otherwise instruct. The Landscape Architect shall certify the dates when in his opinion the Contractor's obligations under this clause have been discharged.

Note:
*Clauses marked * should be completed/deleted as appropriate*

*B The maintenance of the trees, shrubs, grass and other plants after the date of the said certificate will be undertaken by the Employer who will be responsible for the replacement of any trees, shrubs, grass or other plants which are subsequently defective.

Clause 4·0 Payment

Add to clause 4·3

and unless the failures of plants as set out in clause 2·7 are in excess of 10% in which case the amount retained shall be adjusted accordingly.

Clause 6·0 Injury, damage and insurance

Add new clause 6·4

Malicious Damage or Theft (before Practical Completion)

†6·4A All loss or damage arising from any theft or malicious damage prior to practical completion shall be made good by the Contractor at his own expense.

B The Contract sum shall include the provisional sum of £.............. * to be expended as instructed by the Landscape Architect in respect of the cost of all work arising from any theft or malicious damage to the works beyond the control of the Contractor prior to practical completion of the works.

Signed .... date ....
for and on behalf of the Employer

Signed .... date ....
for and on behalf of the Contractor

JCLI

These Supplementary Clauses are issued by the Joint Council for Landscape Industries comprising:-

Landscape Institute
Horticultural Trades Association
British Association Landscape Industries
National Farmers' Union
Institute of Leisure and Amenity Management

Published for the Joint Council for Landscape Industries by the Landscape Institute
6/7 Barnard Mews, London SW11 1QU
and available from RIBA Publications Limited
Finsbury Mission, 39 Moreland Street,
London EC1V 8BB and from RIBA Bookshops

Printed by Duolith Ltd.
Welwyn Garden City, Hertfordshire

Revised October 1991

Appendix E JLCPS General Conditions, Specifications and Schedules of Quantity for the Supply and Delivery of Plants

THE

RECOMMENDED

STANDARD FORM OF TENDER FOR

THE SUPPLY

and DELIVERY

of PLANTS

7th Edition – Published 1992

THIS DOCUMENT HAS BEEN PREPARED AND PUBLISHED

BY THE COMMITTEE ON PLANT SUPPLY AND ESTABLISHMENT (CPSE)

AND APPROVED BY ITS CONSTITUENT BODIES:

Arboricultural Association
Association of County Councils
Association of District Councils
British Association of Landscape Industries
Horticultural Trades Association
Institute of Chartered Foresters
Institute of Leisure & Amenity Management
Landscape Institute
National Farmers Union

IT IS RECOMMENDED FOR USE BY FAX AND BY POST FOR THE PURCHASE OF HARDY NURSERY STOCK

OBTAINABLE FROM: THE SECRETARY, CPSE
c/o THE HORTICULTURAL TRADES ASSOCIATION
19 HIGH STREET, THEALE, READING, BERKS. RG7 5AH Tel: (0734) 303132

JULY/1992

GUIDANCE NOTES TO PURCHASER ON HOW TO PREPARE AND USE THIS FORM

1. PURPOSE

This form is intended for all types of enquiries for the supply and delivery of plants.

2. SPECIFICATION

When describing the type and size of plant required, ensure that:–

(a) Container grown plants are marked by the abbreviation "CG" (unless so marked it will be assumed that plants are either bare-rooted or balled).

(b) The designation where applicable and critical dimensions are stated.
The critical dimensions are:–

Younger trees	Height from ground level – see table below
Standard trees	Circumference of stem – see table below
Container grown nursery stock	Volume of container – see current British Container Growers Standards for container grown plants, garden or amenity grades as appropriate (available free from the Horticultural Trades Association at address on cover of this form).
Other upright growing shrubs and conifers	Height from ground level
Other spreading shrubs & ground cover plants	Diameter

(c) Clear stem heights of standard trees are stated if required to be in excess of 1.8m

(d) Minimum stem diameters are stated if required for those plants so specified in BS 3936 Nursery Stock Part 4. Specification for Forest Trees.

3. QUANTITIES

Quantities of each variety should be totalled so the variety is entered only only in the Schedule of Quantities. It is recommended that large quantities be rounded off.

4. TYPING OF SCHEDULE OF QUANTITIES

Trees, shrubs, roses and herbaceous plants etc., should be listed separately, and in alphabetical order. Type the plant name and form or size high up in the appropriate box so as to allow any substitutes to be named clearly, should they be permitted. Where the schedule of quantities is prepared as a computer print-out, ensure it incorporates all the headings in the schedule of quantities in this form, and is similarly spaced.

5. PACKAGING (See also p. 13 opposite)

Any requirements additional to the specification in The Plant Handling Booklet published by the Committee on Plant Supply and Establishment July 1985 should be clearly stated. It must be recognised that these will be charged for as extras at cost.

6. TENDER FORM

Ensure all the asterisked items on the standard form of tender page are either completed or deleted.

7. NUMBER OF COPIES

It is essential that two copies be sent to each supplier requested to quote, so that one copy may be retained on file. Similarly computerised forms should be sent in duplicate. The reply envelope should be included with the enquiry and indicate to whom to be returned, closing date, and time for submission of tenders.

8. TIME ALLOWED FOR TENDERING

It is essential to allow suppliers a minimum of five clear working days in which to tender, and a period of two weeks is recommended where a sealed tender is required. A fax must be confirmed by hard copy.

9. PRE-INSPECTION

Experience has shown that the best way of relating prices tendered by several Suppliers to the quality of the stock offered is by inspection – this is strongly recommended, and should be strictly by appointment.

10. ORDERING

It is strongly recommended that orders are placed as soon as possible after the closing date so as to ensure supply. (See Form of Tender)

DIMENSIONS OF TREES

Additions to British Standards are indicated in BOLD TYPE

Relevant British Standard	Description	Circumference of stem measured 1.00m from ground level	Height from ground level	Clear stem height from ground level to lowest branch
BS3936 Part 4 Specification for Forest Trees	Age: Seedling (1+0) Seedling (2+0) Undercut Seedling (1u1) Undercut Seedling (1u2) Transplant (1+1) Transplant (1+2)	Does not apply	**15 – 30cm** **30 – 45cm** **45 – 60cm** **60 – 90cm** **90 – 120cm**	Does not apply
BS3936 Nursery stock Part 1. Specification for Trees & Shrubs	Whip (Shall have been previously transplanted at least once in its life, shall not necessarily be staked, and shall be without significant feather growth and without head).	Does not apply	1.20 – 1.50m 1.50 – 1.80m 1.80 – 2.10m 2.10 – 2.50m	Does not apply
BS3936 Nursery stock Part 1. Specification for Trees & Shrubs	Feathered (Shall have been previously transplanted at least once in its life, shall have a defined reasonably straight upright central leader, and a stem furnished with evenly spread and balanced lateral shoots down to near ground level, according to its species).	Does not apply	**1.50 – 1.80m** 1.80 – 2.10m 2.10 – 2.50m 2.50 – 3.00m 3.00 – 3.50m	Does not apply

THE PURCHASER MUST COMPLETE OR DELETE ALL ASTERISKED ITEMS IN THIS FORM AS APPROPRIATE

STANDARD FORM OF TENDER FOR THE SUPPLY AND DELIVERY OF PLANTS

To be completed by the Purchaser

To (Name of Purchaser) ____________________ Ref. No. ____________________

____________________ Contact Name ____________________

Address ____________________ Telephone ____________________

____________________ Fax* ____________________

Project ____________________ Date ____________________

* Delivery to the site at ____________________ which will/will not* be continuously manned; if manned, name of person to contact on site

____________________ ____________________

* The plants will be collected by prior arrangement.

Delivery required (month and year) ____________________

THIS TENDER TO BE RETURNED NOT LATER THAN ____________________ (date and time).

To be completed by the Supplier

SUMMARY OF TENDER

Page No.	£	p	Page No.	£	p	Page No.	£	p
1			B/f:			B/f:		
2			6			10		
3			7					
4			8			Special cond. charge (if any)		
5			9			Delivery charge (if any)		
C/f:			C/f:			TOTAL:		
						VAT additional:		

* **EITHER**

Subject to an order being received within 5 working days of the latest date for submission fo this tender, for all the items quoted on our tender, and subject to our not advising the purchaser to the contrary within 3 working days of receipt of such an order, we undertake to supply/supply and deliver* the plants shown on the attached pages numbered

1 to .... in accordance with the General Conditions for the sum of £..

....

pounds (amount in words) .... pence inclusive of any Value Added Tax due in accordance with the Finance (No. 2) Act of 1975, or subsequent re-enactment.

In the event of an order not being received within 5 working days, this tender remains open for acceptance within a further 4 weeks, subject to the plants being unsold on receipt of order.

If the plants are to be supplied through a contractor whose identity is not known to us at the time of tender, we reserve the right to withdraw this tender within 14 days of the name of the contractor being made known to us.

* **OR**

Subject to the plants being unsold on receipt of order, we undertake to supply/supply and deliver* the plants

shown on the attached pages number 1 to .... in accordance with the General Conditions for the sum

of £..

....

pounds (amount in words) .... pence exclusive of any Value Added Tax due in accordance with the Finance (No. 2) Act 1975 or subsequent re-enactment.

This tender is open for acceptance within 5 weeks from the date for submission of this tender.

If the plants are to be supplied through a contractor whose identity is not known to us at the time of tender, we reserve the right to withdraw this tender within 14 days of the name of the contractor being made known to us.

Date ____________________ Signed ____________________

(For and on behalf of the Supplier)

Tel ____________________ Contact Name ____________________

Fax ____________________ Name of Supplier ____________________

____________________ Ref. No. ____________________

Address ____________________

____

JULY/1992

DIMENSIONS OF TREES (continued)

Relevant British Standard	Description	Circumference of stem measured 1.00m from ground level	Height from ground level	Clear stem height from ground level to lowest branch
BS3936 Nursery stock Part 1. Specification for Trees & Shrubs	Short Standard	Not specified	Not specified	1.00 – 1.20m
	Half Standard	Not specified	1.80 – 2.10m	1.20 – 1.50m
	Extra Light Standard	4 – 6cm	2.10 – 2.50m	1.50 – 1.80m
	Light Standard	6 – 8cm	2.50 – 2.75m	1.50 – 1.80m
	Standard	8 – 10cm	2.75 – 3.00m	1.80m min.
	Tall Standard	8 – 10cm	3.00 – **3.50m**	1.80m min.
	Selected Standard	10 – 12cm	3.00 – **3.50m**	1.80m min.
	Weeping trees – as above but height from ground level does not apply			
BS5236 Recommendations for cultivation and planting of trees in advanced Nursery Stock category	Heavy Standard	12 – 14cm	**3.50 – 4.25m**	1.80m min.
	Extra Heavy Standard	14 – 16cm	**4.25 – 6.00m**	1.80m min.
	Extra Heavy Standard	16 – 18cm	**4.50 – 6.25m**	1.80m min.
	Extra Heavy Standard	18 – 20cm	**4.50 – 6.50m**	1.80m min.
BS4043 Recommendations for Transplanting Semi-Mature Trees	Semi-Mature	20 – 75cm	6.0 – 12.00m	To be specified
		Note: It is recommended that semi-mature trees are chosen by inspection. The relevant British Standard for semi-mature trees is BS4043.		

GENERAL CONDITIONS

1. **GENERAL SPECIFICATION:** The nursery stock must comply in all respects with the current relevant editions of British Standards except for trees in BS3936 Parts 1 and 4 and in BS5236 for which provisions extending those in British Standards are incorporated in this document. All container grown nursery stock is to comply with the current edition of British Container Growers Standards for Container grown plants. (Obtainable from the Horticultural Trades Association at address on cover of this form.)
2. **COUNTRY WHERE GROWN:** The country where grown is to be stated by the Supplier in the Schedule of Quantities. Plants grown in Britain for one growing season or longer shall be considered British grown.
3. **INSPECTION:** The purchaser reserves the right to inspect the stock by appointment, before and/or after placing an order.
4. **PARTIAL ORDERS:** The purchaser reserves the right to accept offers in whole or in part.
5. **PHASED AND/OR MULTISITE DELIVERIES:** This must be stated with the enquiry.
6. **ORDER:** No plants are to be supplied except on receipt of a signed official order. The purchaser is not bound to accept the lowest or any offer.
7. **SUBSTITUTES:** Where stocks are not available on receipt of order substitutes may only be made by mutual agreement.
8. **PLANT HANDLING:** Packaging and transporting shall be as specified in the booklet on Plant Handling published by the Committee on Plant Supply and Establishment July 1985. Plant handling from lifting until delivery on site shall be as recommended in this booklet.
9. **DELIVERY DATES**

   (a) Delivery is to be by the date specified by the purchaser, subject to exceptionally adverse weather or other factors beyond the supplier's control. The supplier is to advise in advance of the week of delivery and for sites not continuously manned to give 48 hours notice of the date and time of delivery. Delivery is to be during normal working hours. Delivery is to be to the one location specified except where otherwise stated, and may be signed for as unexamined.

   (b) In the event of the supplier failing to deliver the whole or part of an order by the date specified, for reasons other than exceptionally adverse weather or other factors beyond the supplier's control, the purchaser and supplier shall agree within 24 hours of that date on a revised delivery date to be made within 5 working days, or agree alternative arrangements. Where cancellation of all or part of the order by the supplier becomes necessary, all agreed costs incurred thereby by the purchaser may be deducted from any monies due, or to become due to suppliers under this order, or shall be recoverable from the supplier by the purchaser as a debt.
10. **COMPLAINTS:** Complaints regarding quantity, quality or condition on delivery must be made in writing within 7 days of receipt of the goods. The purchaser has the right to return, at the supplier's expense, any plants supplied not to specification. This general condition will not vitiate the purchaser's rights to seek consideration should plants display characteristics different to those quoted for, which are undetectable during dormancy, and manifest themselves during the first season.
11. **PAYMENT:** Invoices are due for monthly settlement. Invoices may be rendered for each consignment. Prices quoted are strictly nett and are inclusive of delivery. Value Added Tax will be charged at the rates ruling at the time of delivery.
12. **POSTPONED DELIVERIES:** Postponement by the purchaser of the delivery of any plants after the agreed date shall be only extended within the current planting season.
13. **CANCELLATION:** The supplier has the right on cancellation of whole or part of the order by the purchaser to invoice for the loss.
14. **ARBITRATION:** In the event of a dispute on an order based on these General Conditions, either party may contact one of the subscribing bodies, named on the front cover of this form, who will ask the President of the Institute of Arbitrators to appoint an Arbitrator.
15. **SPECIAL CONDITIONS:** Any special conditions or requirements are to be clearly stated. It must be recognised that these may incur an additional cost.

JULY/1992

SCHEDULE OF QUANTITIES

Page No. ....

Type plant names in the upper space to allow any alternative to be inserted by the supplier in the lower space, should substitutes be permitted.

To: (Name of Purchaser)..

Item No.	Plant Name	Designation & critical dimension (size or volume)	Quantity	Rate £	TOTAL £	Country where grown (to be stated by supplier)
1						
2						
3						
4						
5						
6						
7						
8						
9						
10						
11						
12						
13						
14						
15						

Name of Supplier:

.... Page total carried forward to summary sheet............ £

JULY/1992

SCHEDULE OF QUANTITIES

Page No. ....

Type plant names in the upper space to allow any alternative to be inserted by the supplier in the lower space, should substitutes be permitted.

To: (Name of Purchaser)...

Item No.	Plant Name	Designation & critical dimension (size or volume)	Quantity	Rate £	TOTAL £	Country where grown (to be stated by supplier)
16						
17						
18						
19						
20						
21						
22						
23						
24						
25						
26						
27						
28						
29						
30						

Name of Supplier:

.... Page total carried forward to summary sheet............ £

JULY/1992

Appendix F JCT Standard Form of Tender of Nominated Supplier (Tender TNS/1)

JCT Standard Form of Tender by Nominated Supplier

For use in connection with the Standard Form of Building Contract (SFBC) issued by the Joint Contracts Tribunal, 1980 edition, incorporating Amendments 1 to 9

Job Title: ______________________________
(name and brief location of Works)

____

____

[**a**] To be completed by or on behalf of the Architect/the Contract Administrator.

Employer: [**a**] ______________________________

____

Main Contractor: [**a**] ______________________________
(if known)

____

Tender for: [**a**] ______________________________
(abbreviated description)

____

____

Name of Tenderer: ______________________________

____

To be returned to: [**a**] ______________________________

[**b**] To be completed by the supplier; see also Schedule 1, item 7.

Lump sum price: [**b**] ______________________________

______________________________ *(words)*
and/or Schedule of rates (attached)

Tender TNS/1 (SFBC)

[**c**] By SFBC clause 36·4·9 none of the provisions in the contract of sale can override, modify or affect in any way the provisions incorporated from SFBC clause 36·4 in that contract of sale. Nominated Suppliers should therefore take steps to ensure that their sale conditions do not incorporate any provisions which purport to override, modify or affect in any way the provisions incorporated from SFBC clause 36·4.

1 We confirm that we will be under a contract with the Main Contractor:

·1 to supply the materials or goods described or referred to in **Schedule 1** for the price and/or at the rate set out above; and

·2 in accordance with the other terms set out in that Schedule, as a Nominated Supplier in accordance with the terms of SFBC clause 36·3 to ·5 (as set out in **Schedule 2**) and our conditions of sale in so far as they do not conflict with the terms of SFBC clause 36·3 to ·5[**c**]

provided:

·3 the Architect/the Contract Administrator has issued the relevant nomination instruction (a copy of which has been sent to us by the Architect/the Contract Administrator); and

·4 agreement on delivery between us and the Main Contractor has been reached as recorded in **Schedule 1** Part 6 (see SFBC clause 36·4·3); and

·5 we have thereafter received an order from the Main Contractor accepting this tender.

[**d**] May be completed by or on behalf of the Architect/the Contract Administrator; if not so completed, to be completed by the supplier.

2 We agree that this Tender shall be open for acceptance by an order from the Main Contractor within [**d**] of the date of this Tender. Provided that where the Main Contractor has not been named above we reserve the right to withdraw this Tender within 14 days of having been notified, by or on behalf of the Employer named above, of the name of the Main Contractor.

[**e**] To be struck out by or on behalf of the Architect/the Contract Administrator if no Warranty Agreement is required.

3[**e**] Subject to our right to withdraw this Tender as set out in paragraph 2 we hereby declare that we accept the Warranty Agreement in the terms set out in **Schedule 3** hereto on condition that no provision in that Warranty Agreement shall take effect unless and until

a copy to us of the instruction nominating us,
the order of the Main Contract accepting this Tender, and
a copy of the Warranty Agreement signed by the Employer

have been received by us.

For and on behalf of

Address____

____

Signature____________________Date____________________

Schedule 1

1. Description, quantity and quality of material or goods:

   1A

   1B

   1C

   Note: 1A to be completed by or on behalf of the Architect/the Contract Administrator setting out his requirements. If the supplier is unable to comply with 1A he is to state in 1B what modifications he proposes, and the Architect/the Contract Administrator is to state in 1C if such modifications are acceptable.

2. Access to Works:

   2A

   2B

   Note: 2A to be completed by or on behalf of the Architect/the Contract Administrator. The supplier in 2B **either** *confirms that the access in 2A is acceptable* **or** *states what modifications etc. to the access he requires* **or** *if 2A has not been completed, completes 2B.*

3. Provisions, if any, for returnable packings:

4. Date for Completion of Main Contract (or anticipated Date for Completion if Main Contract not let):

   Note: To be completed by or on behalf of the Architect/the Contract Administrator.

5. Defects Liability Period of the Main Contract months.

   Note: To be completed by or on behalf of the Architect/the Contract Administrator.

6A Anticipated commencement and completion dates for the nominated supply after the necessary approval of drawings (subject to SFBC clause 36·4·3, which provides that delivery shall be commenced, carried out and completed in accordance with a delivery programme to be agreed between the Contractor and supplier including, to the extent agreed, the grounds set out in clause 36·4·3 on which that programme may be varied or if no such programme is agreed in accordance with the reasonable directions of the Contractor):

6B ·1 Supplier's proposed delivery programme to comply with 6A:

·2 If 6B·1 not completed, delivery shall be commenced, carried out and completed in accordance with the reasonable directions of the Contractor.

6C Delivery programme as agreed between the supplier and the Contractor, if different from 6B·1:

Note: 6A to be completed by or on behalf of the Architect/the Contract Administrator. The supplier to complete 6B·1 to take account of 6A; but the completion of 6B·1 is subject to the terms of 6C which may need to be used when the Contractor and supplier are settling item 6.

7. Provisions, if any, for fluctuations in price or rates:

7A

7B

7C

Note: 7A to be completed by or on behalf of the Architect/the Contract Administrator. If the supplier is unable to comply with 7A he is to state in 7B what modifications to the provisions in 7A he requires, and the Architect/the Contract Administrator is to state in 7C if such modifications are acceptable.

8. SFBC clause 25 (Extension of time) applies to the Main Contract without modification except as stated below:

The liquidated and ascertained damages (SFBC clause 24·2·1 and Appendix entry) under the Main Contract are at the rate of:

£__________ per__________.

Note: To be completed by or on behalf of the Architect/the Contract Administrator.

9. Contract of sale with Contractor to be executed under hand/as a deed.

Note: Alternative not to be used to be deleted by the supplier subject to agreement on the method of execution of the sale contract with the Main Contractor.

Schedule 2

JCT Standard Form of Building Contract

Clauses 36·3 to 36·5 provide as follows:

Ascertainment of costs to be set against prime cost sum

36·3 ·1 For the purposes of clause 30·6·2·8 the amounts 'properly chargeable to the Employer in accordance with the nomination instruction of the Architect/the Contract Administrator' shall include the total amount paid or payable in respect of the materials or goods less any discount other than the discount referred to in clause 36·4·4, properly so chargeable to the Employer and shall include where applicable:

·1·1 any tax (other than any value added tax which is treated, or is capable of being treated, as input tax (as referred to in the Finance Act 1972) by the Contractor) or duty not otherwise recoverable under this Contract by whomsoever payable which is payable under or by virtue of any Act of Parliament on the import, purchase, sale, appropriation, processing, alteration, adapting for sale or use of the materials or goods to be supplied; and

·1·2 the net cost of appropriate packing, carriage and delivery after allowing for any credit for return of any packing to the supplier; and

·1·3 the amount of any price adjustment properly paid or payable to, or allowed or allowable by the supplier less any discount other than a cash discount for payment in full within 30 days of the end of the month during which delivery is made.

·2 Where in the opinion of the Architect/the Contract Administrator the Contractor properly incurs expense, which would not be reimbursed under clause 36·3·1 or otherwise under this Contract, in obtaining the materials or goods from the Nominated Supplier such expense shall be added to the Contract Sum.

Sale contract provisions – Architect's/Contract Administrator's right to nominate supplier

36·4 Save where the Architect/the Contract Administrator and the Contractor shall otherwise agree, the Architect/the Contract Administrator shall only nominate as a supplier a person who will enter into a contract of sale with the Contractor which provides, inter alia:

·1 that the materials or goods to be supplied shall be of the quality and standard specified provided that where and to the extent that approval of the quality of materials or of the standards of workmanship is a matter for the opinion of the Architect/the Contract Administrator, such quality and standards shall be to the reasonable satisfaction of the Architect/the Contract Administrator;

·2 that the Nominated Supplier shall make good by replacement or otherwise any defects in the materials or goods supplied which appear up to and including the last day of the Defects Liability Period under this Contract and shall bear any expenses reasonably incurred by the Contractor as a direct consequence of such defects provided that:

·2·1 where the materials or goods have been used or fixed such defects are not such that reasonable examination by the Contractor ought to have revealed them before using or fixing;

·2·2 such defects are due solely to defective workmanship or material in the materials or goods supplied and shall not have been caused by improper storage by the Contractor or by misuse or by any act or neglect of either the Contractor, the Architect/the Contract Administrator or the Employer or by any person or persons for whom they may be responsible or by any other person for whom the Nominated Supplier is not responsible;

·3 that delivery of the materials or goods supplied shall be commenced, carried out and completed in accordance with a delivery programme to be agreed between the Contractor and the Nominated Supplier including, to the extent agreed, the following grounds on which that programme may be varied:

force majeure; or

civil commotion, local combination of workmen, strike or lock-out; or

any instruction of the Architect/the Contract Administrator under clause 13·2 (Variations) or clause 13·3 (provisional sums); or

failure of the Architect/the Contract Administrator to supply to the Nominated Supplier within due time any necessary information for which he has specifically applied in writing on a date which was neither unreasonably distant from nor unreasonably close to the date on which it was necessary for him to receive the same; or

exceptionally adverse weather conditions

or, if no such programme is agreed, delivery shall be commenced, carried out and completed in accordance with the reasonable directions of the Contractor.

·4 that the Nominated Supplier shall allow the Contractor a discount for cash of 5 per cent on all payments if the Contractor makes payment in full within 30 days of the end of the month during which delivery is made;

·5 that the Nominated Supplier shall not be obliged to make any delivery of materials or goods (except any which may have been paid for in full less only any discount for cash) after the determination (for any reason) of the Contractor's employment under this Contract;

·6 that full discharge by the Contractor in respect of payments for materials or goods supplied by the Nominated Supplier shall be effected within 30 days of the end of the month during which delivery is made less only a discount for cash of 5 per cent if so paid;

·7 that the ownership of materials or goods shall pass to the Contractor upon delivery by the Nominated Supplier to or to the order of the Contractor, whether or not payment has been made in full;

·8·1 that in any dispute or difference between the Contractor and the Nominated Supplier which is referred to arbitration the Contractor and the Nominated Supplier agree and consent pursuant to Sections 1(3)(a) and 2(1)(b) of the Arbitration Act 1979 that either the Contractor or the Nominated Supplier

– may appeal to the High Court on any question of law arising out of an award made in the arbitration and

– may apply to the High Court to determine any question of law arising in the course of the arbitration;

and that the Contractor and the Nominated Supplier agree that the High Court should have jurisdiction to determine any such questions of law;

·8·2 that if any dispute or difference between the Contractor and the Nominated Supplier raises issues which are substantially the same as or are connected with issues raised in a related dispute between the Employer and the Contractor under this contract then, where * clauses 41·2·1 and 41·2·2 apply, such dispute or difference shall be referred to the Arbitrator to be appointed pursuant to clause 41; that the Arbitrator shall have power to make such directions and all necessary awards in the same way as if the procedure of the High Court as to joining one or more defendants or joining co-defendants or third parties was available to the parties; that the agreement and consent referred to in clause 36·4·8·1 on appeals or applications to the High Court on any question of law shall apply to any question of law arising out of the awards of such arbitrator in respect of all related disputes referred to him or arising in the course of the reference of all the related disputes referred to him; and that in any case, subject to the agreement referred to in clause 36·4·8·1, the award of such Arbitrator shall be final and binding on the parties.

·9 that no provision in the contract of sale shall override, modify or affect in any way whatsoever the provisions in the contract of sale which are included therein to give effect to clauses 36·4·1 to 36·4·9 inclusive.

Contract of sale – restriction, limitation or exclusion of liability

36·5 ·1 Subject to clauses 36·5·2 and 36·5·3, where the said contract of sale between the Contractor and the Nominated Supplier in any way restricts, limits or excludes the liability of the Nominated Supplier to the Contractor in respect of materials or goods supplied or to be supplied, and the Architect/the Contract Administrator has specifically approved in writing the said restrictions, limitations or exclusions, the liability of the Contractor to the Employer in respect of the said materials or goods shall be restricted, limited or excluded to the same extent.

·2 The Contractor shall not be obliged to enter into a contract with the Nominated Supplier until the Architect/the Contract Administrator has specifically approved in writing the said restrictions, limitations or exclusions.

·3 Nothing in clause 36·5 shall be construed as enabling the Architect/the Contract Administrator to nominate a supplier otherwise than in accordance with the provisions stated in clause 36·4.

*The Architect/The Contract Administrator should state whether in the Appendix to the SFBC the words 'clauses 41·2·1 and 41·2·2' have been deleted; if so then clause 36·4·8·2 will not apply to the Nominated Supplier.

Warranty TNS/2 (SFBC)

Schedule 3: Warranty Agreement by a Nominated Supplier

To the Employer: ____________________

____

named in our Tender dated ____________________

For ____________________
(abbreviated description of goods/materials)

____

To be supplied to: ____________________
(job title)

____

1 Subject to the conditions stated in the above mentioned Tender (that no provision in this Warranty Agreement shall take effect unless and until the instruction nominating us, the order of the Main Contractor accepting the Tender and a copy of this Warranty Agreement signed by the Employer have been received by us) WE WARRANT in consideration of our being nominated in respect of the supply of the goods and/or materials to be supplied by us as a Nominated Supplier under the Standard Form of Building Contract referred to in the Tender and in accordance with the description, quantity and quality of the materials or goods and with the other terms and details set out in the Tender ('the supply') that:

1·1 We have exercised and will exercise all reasonable skill and care in:

1·1 ·1 the design of the supply insofar as the supply has been or will be designed by us; and

·2 the selection of materials and goods for the supply insofar as such supply has been or will be selected by us; and

·3 the satisfaction of any performance specification or requirement insofar as such performance specification or requirement is included or referred to in the Tender as part of the description of the supply.

1·2 We will:

1·2 ·1 save insofar as we are delayed by:

·1 force majeure; or

·2 civil commotion, local combination of workmen, strike or lock-out; or

·3 any instruction of the Architect/the Contract Administrator under SFBC clause 13·2 (Variations) or clause 13·3 (provisional sums); or

Pages 1 to 6 comprising Tender and Schedules 1 and 2 are issued in a separate pad, TNS/1 (SFBC).

Warranty TNS/2 (SFBC)

1·2 ·1 *continued*

·4 failure of the Architect/the Contract Administrator to supply to us within due time any necessary information for which we have specifically applied in writing on a date which was neither unreasonably distant from nor unreasonably close to the date on which it was necessary for us to receive the same

so supply the Architect/the Contract Administrator with such information as the Architect/the Contract Administrator may reasonably require; and

·2 so supply the Contractor with such information as the Contractor may reasonably require in accordance with the arrangements in our contract of sale with the Contractor; and

·3 so commence and complete delivery of the supply in accordance with the arrangements in our contract of sale with the Contractor

that the Contractor shall not become entitled to an extension of time under SFBC clauses 25·4·6 or 25·4·7 of the Main Contract Conditions nor become entitled to be paid for direct loss and/or expense ascertained under SFBC clause 26·1 for the matters referred to in clause 26·2·1 of the Main Contract Conditions; and we will indemnify you to the extent but not further or otherwise that the Architect/the Contract Administrator is obliged to give an extension of time so that the Employer is unable to recover damages under the Main Contract for delays in completion, and/or pay an amount in respect of direct loss and/or expense as aforesaid because of any failure by us under clause 1·2·1 or 1·2·2 hereof.

2 We have noted the amount of the liquidated and ascertained damages under the Main Contract, as stated in TNS/1 Schedule 1, item 8.

3 Nothing in the Tender is intended to or shall exclude or limit our liability for breach of the warranties set out above.

4·1 In case any dispute or difference shall arise between the Employer or the Architect/the Contract Administrator on his behalf and ourselves as to the construction of this Agreement or as to any matter or thing of whatsoever nature arising out of this Agreement or in connection therewith then such dispute or difference shall be and is hereby referred to arbitration. When we or the Employer require such dispute or difference to be referred to arbitration we or the Employer shall given written notice to the other to such effect and such dispute or difference shall be referred to the arbitration and final decision of a person to be agreed between the parties as the Arbitrator, or, upon failure so to agree within 14 days after the date of the aforesaid written notice, of a person to be appointed as the Arbitrator on the request of either ourselves or the Employer by the person named in the Appendix to the Standard Form of Building Contract referred to in the Tender.

4·2 ·1 Provided that if the dispute or difference to be referred to arbitration under this Agreement raises issues which are substantially the same as or connected with the issues raised in a related dispute between the Employer and the Contractor under the Main Contract or between a Nominated Sub-Contractor and the Contractor under Sub-Contract NSC/4 or NSC/4a or between the Employer and any other Nominated Supplier, and if the related dispute has also been referred for determination to an Arbitrator, the Employer and ourselves hereby agree that the dispute or difference under this Agreement shall be referred to the Arbitrator appointed to determine the related dispute; and the JCT Arbitration Rules applicable to the related dispute shall apply to the dispute under this Agreement; and such Arbitrator shall have power to make such directions and all necessary awards in the same way as if the procedure of the High Court as to joining one or more of the defendants or joining co-defendants or third parties was available to the parties and to him; and the agreement of consent referred to in paragraph 4·6 on appeals or applications to the High Court on any question of law shall apply to any question of law arising out of the awards of such Arbitrator in respect of all related disputes referred to him or arising in the course of the reference of all the related disputes referred to him.

·2 Save that the Employer or ourselves may require the dispute or difference under this Agreement to be referred to a different Arbitrator (to be appointed under this Agreement) if either of us reasonably considers that the Arbitrator appointed to determine the related dispute is not properly qualified to determine the dispute or difference under this Agreement.

·3 Paragraphs 4·2·1 and 4·2·2 hereof shall apply unless in the Appendix to the Standard Form of Building Contract referred to in the Tender the words 'clause 41·2·1 and 41·2·2 apply' have been deleted.

4·3 Such reference shall not be opened until after Practical Completion or alleged Practical Completion of the Main Contract Works or termination or alleged termination of the Contractor's employment under the Main Contract or abandonment of the Main Contract Works, unless with the written consent of the Employer or the Architect/the Contract Administrator on his behalf and ourselves.

4·4 Subject to paragraph 4·5 the award of such Arbitrator shall be final and binding on the parties.

4·5 The parties hereby agree and consent pursuant to Sections 1(3) and 2(1) (b) of the Arbitration Act, 1979, that either party

·1 may appeal to the High Court on any question of law arising out of an award made in any arbitration under this Arbitration Agreement; and

·2 may apply to the High Court to determine any question of law arising in the course of the reference;

and the parties agree that the High Court should have jurisdiction to determine any such question of law.

4·6 Whatever the nationality, residence or domicile of ourselves or the Employer, the Contractor, any sub-contractor or supplier or the Arbitrator, and wherever the Works or any part thereof are situated, the law of England shall be the proper law of this Warranty and in particular (but not so as to derogate from the generality of the foregoing) the provisions of the Arbitration Acts 1950 (notwithstanding anything in S.34 thereof) to 1979 shall apply to any arbitration under this Contract wherever the same, or any part of it, shall be conducted.[*]

4·7 If before his final award the Arbitrator dies or otherwise ceases to act as the Arbitrator, the Employer and ourselves shall forthwith appoint a further Arbitrator, or, upon failure so to appoint within 14 days of any such death or cessation, then either the Employer or ourselves may request the person named in the Appendix to the Standard Form of Building Contract referred to in the Tender to appoint such further Arbitrator. Provided that no such further Arbitrator shall be entitled to disregard any direction of the previous Arbitrator or to vary or revise any award of the previous Arbitrator except to the extent that the previous Arbitrator had power so to do under the JCT Arbitration Rules and/or with the agreement of the parties and/or by the operation of law.

4·8 The arbitration shall be conducted in accordance with the 'JCT Arbitration Rules' current at the date of the Tender. Provided that if any amendments to the Rules so current have been issued by the Joint Contracts Tribunal after the aforesaid date the Employer and Supplier may, by a joint notice in writing to the Arbitrator, state that they wish the arbitration to be conducted in accordance with the JCT Arbitration Rules as so amended.[†]

[††]Signature of or on behalf of the Supplier: ______________________

[††]Signature of or on behalf of the Employer: ______________________

---

[*] Where the parties do not wish the proper law of the Warranty to be the law of England appropriate amendments to paragraph 4·7 should be made. Where the Works are situated in Scotland then the forms issued by the Scottish Building Contract Committee which contain Scots proper law and arbitration provisions are the appropriate documents. It should be noted that the provisions of the Arbitration Acts 1950 to 1979 do not apply to arbitrations conducted in Scotland.

[†] The JCT Arbitration Rules contain stricter time limits than those prescribed by some arbitration rules or those frequently observed in practice. The parties should note that a failure by a party or the agent of a party to comply with the time limits incorporated in these Rules may have adverse consequences.

[††] If the Warranty Agreement is to be executed as a deed advice should be sought on the correct method of execution.

Appendix G NJCC Code of Procedure for Single-stage Selective Tendering

CONTENTS *page*

Note: The Code applies in the whole of the United Kingdom, but where Scottish practice differs, suitable recomendations are printed in *italics.*

1.0 FOREWORD

1.1 This Code has been prepared for all who commission building work, whether they be private clients or public authorities, and envisages the use of traditional single stage tendering by a selected list of tenderers. It is the prerogative of the employer to decide the method of tendering to be adopted, but the National Joint Consultative Committee for Building and the Department of the Environment believe that single stage selective tendering will be found the most appropriate method of obtaining tenders for the majority of building contracts. On contracts where it is desired to secure the early involvement of the contractor before the development of the design is completed two stage tendering procedures may be adopted. These procedures are detailed in a companion NJCC Code of Procedure for Two Stage Selective Tendering. Nothing in this Code should be taken to suggest that the employer is obliged to accept the lowest or any tender, although if the procedure advocated in the Code is followed, the successful tenderer will normally be the one offering the lowest price.

1.2 Consequent upon entry into the European Economic Community tenders for public sector construction contracts above a specified value must be invited and contracts awarded in accordance with Council Directive 71/305/EEC (as amended by 78/669/EEC) and 72/277/EEC. In these instances the provisions of this Code will be qualified by the supplementary tendering procedures specified in the Directive. Guidance on the operation of the procedures laid down in the Directives is given in DoE Circulars 59/73 (obtainable from HMSO) and 102/76 and 67/78 (obtainable from DoE).

*1.2 *In Scotland guidance on the operation of the procedures laid down in the EEC Directives is given in SDD Circular 45/73, SOFD Circulars 43/76 and 24/78, and SDD Circular 2/82, all obtainable from the Scottish Development Department.*

1.3 This Code is complementary to other NJCC publications (listed opposite).

1.4 Throughout the Code the word 'Architect' is used to mean Architect [1] or Contract Administrator and the word 'employer' to mean building owner or employing body in the sense that the terms are used in the standard forms of building contract.

---

[1] The term 'Architect' applies only to any person entitled to use the name 'Architect' under and in accordance with the Architects Registration Acts 1931 to 1969.

2.0 INTRODUCTION

2.1 The general principles of the Code are fully supported by the Department of the Environment and take into account the relevant recommendations of the National Economic Development Office (NEDO) report 'The Public Client and the Construction Industries (1975)', the report 'The Placing and Management of Contracts for Building and Civil Engineering Work' (The Banwell Report), and of the 'Action on the Banwell Report'.

2.2 In order that its detailed recommendations may be most widely applicable, this Code assumes, in the contract which will follow tendering, the use of the Standard Form of Building Contract (Private and Local Authorities editions), the Intermediate Form of Building Contract or the Agreement for Minor Building Works, all issued by the Joint Contracts Tribunal. This assumption in no way precludes the use of the Code where other forms of contract are employed, though certain details may require some modification.

2.3 The contractor's tender is the price for which he offers to carry out and complete, in accordance with the conditions of the contract, the works shown on the drawings and described in the bill(s) of quantities and/or specifications and schedules.

2.4 Good tendering procedure demands that the contractor's tendered price should not be altered without justification. Difficulties have arisen when an examination of the priced bill(s) revealed errors or a discrepancy between these prices and the tender figure. The Code lays down, in Section 6, alternative methods of dealing with this situation before the contractor's offer is accepted. The first alternative does not permit the correction of errors in priced bill(s) of quantities, the second does.

The choice must be made before contractors are invited to tender.

2.5 It should be a condition of tender that obvious errors in pricing or significant errors in arithmetic discovered before the acceptance of the contractor's tender should be dealt with in accordance with one of the alternatives in Section 6 of this Code.

3.0 THE LIST OF TENDERERS

3.1 Once it has been decided that a contractor is to be selected by competitive tender a short list of suitable tenderers should be drawn up either from the employer's approved list of contractors or from an *ad hoc* list of contractors of established skill, integrity, responsibility and proven competence for work of the character and size contemplated. It is recommended that the number of tenderers invited should be limited to a maximum of six.

When the list has been settled, one or two further names should be appended in order that they may replace any firms on the list that do not accept the preliminary invitation.

3.2 When selecting the short list the following are among the points which should be considered:

.1 the firm's financial standing and record;

.2 whether the firm has had recent experience of building at the required rate of completion over a comparable contract period;

.3 the firm's general experience and reputation in the area in question;

.4 whether the management structure of the firm is adequate for the type of contract envisaged; and

.5 whether the firm will have adequate capacity at the relevant time

3.3 Approved lists should be reviewed periodically to exclude firms whose performance has been unsatisfactory, and to allow the introduction of suitable additional firms.

3.4 The object of selection is to make a list of firms, any one of which could be entrusted with the job. If this is achieved, then the final choice of contractor will be simple – the firm offering the lowest tender. Only the most exceptional cases justify departure from this general recommendation.

4.0 TENDERING PROCEDURE

4.1 Preliminary Enquiry

.1 In order that contractors may be able to decide whether they will tender, and to anticipate demands on their tendering staff, each firm should be sent, and should reply promptly to, a preliminary invitation to tender as illustrated in Appendix A – 'Preliminary Enquiry for Invitation to Tender'.

It is essential that all the details itemised in Appendix A are stated in the preliminary enquiry. The omission of relevant information may seriously impede contractors in deciding whether or not to tender.

.2 It is suggested that an appropriate period of time between the preliminary enquiry and the despatch of tender documents is 4 to 6 weeks, although in some instances a period of 3 months would not be unreasonable. In cases beyond 3 months the preliminary invitation should be confirmed a month before tenders are invited.

.3 Once a contractor has signified initial agreement to tender it is in the best interests of all parties that such acceptance should be honoured. If in exceptional circumstances a contractor has to withdraw his acceptance he should give notice of this intention before the issue of tender documents. If for any reason this is not possible, notice should be given not later than two working days after receipt of the tender documents.

.4 After the latest date for the acceptance of the preliminary invitation the final short list of tenders will be selected, and firms which notified their willingness to tender, but are not included in the tender list, should be promptly informed.

4.2 **The Tender Documents**

.1 On the day stated in the preliminary invitation all tender documents should be available for collection or despatched to the tenderers by first class post. See Appendix B for suggested 'Formal Invitation to Tender' and Appendix C for suggested 'Form of Tender'.

.2 The conditions of tendering should be absolutely clear so that all tenders are submitted on the same basis. The contract period should be specified in the tender documents and tenderers should be required to make offers based on the same period of completion in order to limit competition to price only. Tenders for alternative periods of completion should only be admitted by invitation at the time of receipt by tenderer of the tender documents and in the event of alternative tenders being sanctioned they must all be submitted at the same time.

.3 The NJCC strongly recommends the use of the standard forms of building contract in unamended form.

Although it is recognised that the terms of the standard forms are not mandatory and that the employer and his professional advisers may alter them at their discretion, the NJCC believes that alterations to the standard forms impede the efforts being made towards achieving greater standardisation of building procedures. The NJCC is firmly of the opinion that if alterations to the standard forms have to be made it is essential in the interests of good practice and of economic building that they be kept to an absolute minimum. They should not be undertaken without serious prior consideration and should then be drafted by a person competent to ensure that all consequential alterations to other clauses are made (see NJCC Procedure Note 2).

The tenderer's attention should be specifically drawn in the Preliminary Invitation to Tender (see Appendix A) to any alterations to be made to the standard form of contract and, where appropriate, reasons should be given, so that the implications of such amendments may be considered by the tenderers prior to acceptance of the invitiation to tender, and so minimise the risk of subsequent queries at tender stage which may result in an extension of the tender period (see clause 4.4).

4.3 **Time for Tendering**

.1 The time allowed for the preparation of tenders should be determined in relation to the size and complexity of the job. Inadequate tendering time can lead to mistakes and the client may not obtain the most competitive prices. A minimum of four working weeks (20 working days) should normally be allowed. Major projects, plan and specification tenders, works having a significant non-nominated specialist content or other special circumstances may well require a longer period. The tender period must be sufficient to enable the tenderer to obtain competitive quotations for the supply of materials and for the execution of works to be sub-let. The latest time for submission should be specified as an hour of a day and should be chosen to allow as short a time as possible to elapse before opening the tenders. Tenders received after time should not be admitted to the competition.

.2 If bills of quantities are issued to tenderers in sections, it is important that the time for tendering should be calculated from the date of issue of the last section.

4.4 **Qualified Tenders**

.1 For fair competitive tendering it is essential that the tenders submitted by each tenderer be based on identical tender documents and that the tenderers should not attempt to vary that basis by qualifying their tenders (see clause 4.4.2).

.2 If a tenderer considers that any of the tender documents are deficient in any respect and require clarification, or contain unacceptable alterations to the standard form of building contract not previously set out in the Preliminary Invitation to Tender (see Appendix A) he should inform the issuing authority or the architect (with a copy to the quantity surveyor) as soon as possible and preferably not less than ten days before the tenders are due. If it is decided to amend the documents the issuing authority or architect should inform all tenderers and extend the time for tendering if necessary.

.3 A tenderer who otherwise submits a qualified tender should be given the opportunity to withdraw the qualifications without amendment to his tender; if he fails to do so his whole tender should be rejected if it is considered that such qualifications afford the tenderer an unfair advantage over other tenderers.

4.5 **Withdrawal of Tender before Acceptance**

.1 Under English law a tender may be withdrawn at any time before its acceptance.

*.1 *Under Scots law, if the provision for the tender to be withdrawn at any time before its acceptance is to be available, the words 'unless previously withdrawn' should be inserted in the tender after the words 'this tender remains open for consideration. . .' (Note: this clause is not, however, contained in tenders issued in Scotland by the Department of the Environment) (See Appendix C for suggested 'Form of Tender').*

5.0 ASSESSING TENDERS AND NOTIFYING RESULTS

5.1 Tenders should be opened as soon as possible after the time for receipt of tenders.

5.2 The lowest tenderer should be asked to submit his priced bill(s) of quantities as soon as possible and in any case within four working days.

*5.2 *The priced bill(s) of quantities, contained in a sealed, separate envelope, endorsed with the tenderer's name, should be submitted at the same time as the tender.*

5.3 It is important that all but the lowest three tenderers should be informed immediately that their tenders have been unsuccessful, as this information is critical in relation to a contractor's strategic tender planning. In order to serve the employer's interests in the event of the lowest tenderer withdrawing his offer, the second and third lowest tenderers should be informed that their tenders were not the most favourable received but that they will be approached again if it is decided to give further consideration to their offers. They should subsequently be notified at once when a decision to accept a tender has been taken.

*5.3 *The envelope containing the bill(s) of the lowest tenderer should be opened and the bill(s) examined. Alternatively the lowest three tenderers should be advised that their offers are under consideration and the envelopes containing the bill(s) of the lowest three tenderers should be opened and the bill(s) examined. Tenderers whose offers are being rejected should be informed immediately or as early as possible.*

5.4 Once the contract has been let every tenderer should be promptly supplied with a list of tender prices.

*5.4 *Once the contract has been let every tenderer should be supplied with a list of the firms who tendered (in alphabetical order) and list of the tender prices (in ascending order of value).*

6.0 EXAMINATION AND ADJUSTMENT OF THE PRICED BILL(S)

6.1 The examination of the priced bill(s) of quantities supporting the tender under consideration should be made by the quantity surveyor who should treat the document as confidential; on no account should any details of the tenderer's pricing be disclosed to any person, other than the architect or other appropriate consultant, except with the express permission of the tenderer.

6.2 The object of examining priced bills is to detect errors in computation of the tender. If the quantity surveyor finds such errors, he should report them to the architect who, in conjunction with the employer, will determine the action to be taken under whichever is appropriate of the alternatives set out below and referred to in the Formal Invitation to Tender and the Form of Tender. (See Appendices B and C.)

6.3 **Alternative 1**

.1 The tenderer should be given details of such errors and afforded an opportunity of confirming or withdrawing his offer. If the tenderer withdraws, the priced bill(s) of the second lowest should be examined, and if necessary this tenderer be given a similar opportunity.

.2 An endorsement should be added to the priced bill(s) indicating that all rates or prices (excluding preliminary items, contingencies, prime cost and provisional sums) inserted therein by the tenderer are to be considered as reduced or increased in the same proportion as the corrected total of priced items exceeds or falls short of such items. This endorsement should be signed by both parties to the contract.

6.4 **Alternative 2**

.1 The tenderer should be given an opportunity of confirming his offer or of amending it to correct genuine errors. Should he elect to amend his offer and the revised tender is no longer the lowest, the offer of the firm now lowest in the competition should be examined.

*.1 *The lowest tenderer or each of the lowest three tenderers should be given an opportunity of confirming his offer or amending it to correct genuine error.*

---

* Applicable in Scotland only.

6.4 .2 If the tenderer elects not to amend his offer, an endorsement will be required as in 6.3.2. If the tenderer does amend his tender figure, and possibly certain of the rates in his bill(s), he should either be allowed access to his original tender to insert the correct details and to initial them or be required to confirm all the alterations in a letter. If in the latter case his revised tender is eventually accepted, the letter should be conjoined with the acceptance and the amended tender figure and the rates in it substituted for those in the original tender.

6.5 When a tender is found to be free of error, or the tenderer is prepared to stand by his tender in spite of error, or a tender on amendment is still the lowest, this should be recommended to the employer for acceptance.

7.0 NEGOTIATED REDUCTION OF TENDER

7.1 Should the tender under consideration exceed the employer's budget the recommended procedure is for a price to be negotiated with this tenderer. The basis of negotiations and any agreements made should be fully documented.

7.2 Only when these negotiations fail should negotiations proceed with the next lowest tenderer. If these negotiations also fail, similar action may be taken with the third lowest tenderer.

7.3 If all these negotiations fail, new tenders may be called for.

8.0 POST TENDER PERIOD

Although the period between the appointment of a contractor and the commencement of work on site does not strictly fall within the scope of tendering procedure, action taken within this period is so critical to the successful outcome of any project as to justify comment in this Code. The NJCC recommends that a due period, not exceeding two months, be allowed for thorough project planning and for the contractor to organise his resources. Undue haste to make a physical start on site may result in extensive and costly variations which can lead to prolongation and not reduction of the total construction period. Regard also should be had, however, to the fact that unnecessary delay in achieving a start on site may involve the employer in extra costs whether or not the contract is based on variation of price conditions. These points should be borne in mind when determining the anticipated date for possession of the site. (See Appendix A.)

Further reading

Clamp, H. (1988) *The Shorter Forms of Building Contract*. BSP Professional Books, London.
Parris, J. (1982) *The Standard Form of Building Contract*. BSP Professional Books, London.
Turner, D. F. (1985) *Building Subcontract Forms*. BSP Professional Books, London.
Sayers, P. (1991) *Grounds Maintenance: a contractors' guide to competitive tendering*. E & F.N. Spon, London.

Index

Page numbers appearing in **bold** refer to figures and page numbers appearing in *italic* refer to tables.